D0046629

ROLE OF FATS IN FOOD AND NUTRITION

ROLE OF FATS IN FOOD AND NUTRITION

M. I. GURR

National Institute for Research in Dairying,
Shinfield, Reading, UK

ELSEVIER APPLIED SCIENCE PUBLISHERS
LONDON and NEW YORK

ELSEVIER APPLIED SCIENCE PUBLISHERS LTD
Ripple Road, Barking, Essex, England

Sole Distributor in the USA and Canada
ELSEVIER SCIENCE PUBLISHING CO., INC.
52 Vanderbilt Avenue, New York, NY 10017, USA

British Library Cataloguing in Publication Data

Gurr, M. I.
Role of fats in food and nutrition.
1. Nutrition 2. Fat
I. Title
613.2′8 TX553.F3

ISBN 0–85334–298–9

WITH 24 TABLES AND 32 ILLUSTRATIONS

Photoset in Malta by Interprint Ltd
Printed in Great Britain by Galliard (Printers) Ltd, Great Yarmouth

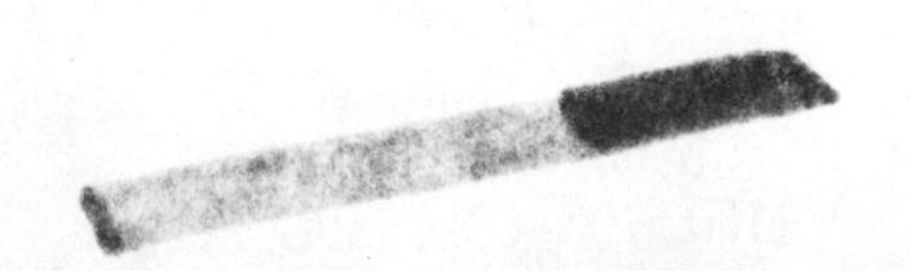

Preface

In countries where food is abundant, it is eaten by most people mainly to satisfy appetite rather than hunger. Eating has become a social activity that gives great pleasure; not a mere matter of staying alive. It is no coincidence that, with few exceptions, the proportion of fat in the diet reflects the affluence of a society and that in countries that develop from relative poverty into affluence, the proportion of dietary fat relative to carbohydrate increases noticeably. It is unlikely that most people in developed countries would be prepared to revert to a diet that is common in many Third World nations.

There is currently more public interest in nutrition than at any time since the Second World War and of the issues being debated in the media and in the pages of scientific journals, the questions 'how much' and 'what sort of fat?' are probably of most widespread interest.

One motivation for writing this book, therefore, was to assess, under one cover, the many attributes of dietary fats that contribute to the nutrition of man in what I hope is a readable form. An important aspect of my treatment of the subject has been to put the nutrition of fats in the context of the nutrition of other food components, since a common myth has been that one can discuss the 'good' or 'bad' aspect of one type of food entirely outwith the context of the diet as a whole. We should not talk about a 'good' or a 'bad' food, only about a 'good' or a 'bad' diet. It is becoming the practice for many nutritionists to regard the old concepts of a 'balanced diet' as passé, but it is my belief that these concepts are more than ever necessary.

There are many works on the modern library bookshelf that deal with the subject of dietary fats. Many are multi-author volumes, some of them emanating from scientific symposia, and reference to them will be found in the bibliographies at the end of each chapter in this book. In these works, the treatment of the subject becomes rather disjointed and it is difficult for the reader to relate one aspect of fat nutrition to another. Others are monographs that tend to concentrate on one or two particular aspects of the subject while others are more in the nature of religious tracts expounding the merits of one point of view or one type of dietary regime.

In this book I have tried as far as possible to cover the whole range of the subject from the chemistry of fats, their occurrence and importance in living things and in man's food, their mode of digestion and absorption and their metabolism in the body. From there it was a natural step to describing their role in maintaining good health as well as in contributing in various ways to ill health. Other books have dealt, in an excellent manner, with fats in the context of food processing, but it has been important here to discuss processing in as far as it influences the nutritional aspects of lipids. As the proportion of our food that is processed in some way or other continues to rise, and as new manufactured foods continue to appear, it is important to look into the future to determine how fat nutrition could be influenced for the benefit of man.

I have tried to walk a tight-rope between breadth of coverage and detail of treatment on the one hand and readability for the non-specialist (with the danger of oversimplification) on the other. I can only hope that I have succeeded in these aims and that readers with diverse backgrounds will find this a useful addition to their bookshelves.

M. I. GURR

Contents

PART I

NATURE, OCCURRENCE AND CHARACTERIZATION OF BIOLOGICALLY IMPORTANT FATS

Chapter 1

What are Fats?

To begin at the beginning: What are fats? Although there is no precise definition, the term 'fat' is generally applied to those foods or components of foods that are clearly fatty in nature, greasy in texture and immiscible with water. Familiar examples are butter and the fatty parts of meats. Fats are thought of as solid in texture as distinct from oils which are liquid at ambient temperatures. Chemically, however, there is little distinction between a fat and an oil since the substances that the layman thinks of as edible fats and oils are all composed predominantly of esters of glycerol with fatty acids. These are called triacylglycerols and are chemically quite distinct from the oils used in the petroleum industry which are hydrocarbons. The melting points of triacylglycerols are largely determined by the degree of unsaturation and chain length of the fatty acids and there will be a more detailed discussion of their chemistry in later chapters.

Chemists use the more general term 'lipid' to describe a chemically varied group of fatty substances, which have in common the property of being insoluble in water, but soluble in solvents such as chloroform, hydrocarbons, alcohols or ethers. This definition includes a far wider range of chemical substances than simply the triacylglycerols. Since there are many fatty substances that are biologically important, the term 'fat' will be used as synonymous with 'lipid' throughout this book to include all fat-soluble substances important in food and nutrition. The term 'lipid', though it may be used on occasions, generally will be avoided as the monograph is not aimed primarily at biochemists but at a much wider readership.

This book is concerned only with human foods although from time to time reference will be made to fats in the diets of farm animals when it is relevant to the composition of human food derived from those animals.

'Nutrition' can be defined very broadly as the study of food and drink in all its aspects. Yet nutrition should not be regarded simply as the needs of the whole person for food, at a rather gross or superficial level. Nutrition is also concerned with the needs of individual body cells for nutrients to maintain their proper physiological function. Because the role of individual fats can be understood only in terms of their contribution to cellular function, a part of this book will necessarily be devoted to simple descriptions of fat metabolism.

The human body and the bodies of other animals are made up of proteins, carbohydrates, fats, minerals and water. Many of the fatty components can be made within the body itself from carbohydrates consumed in the diet. Part of the body fat component, however, may be derived directly from fats in the diet and indeed some kinds of fats that are necessary for life cannot be made by the body itself and have to be supplied in the diet as will be described later. Fats in the diet may come from the body fats of other animals eaten by man or from plants. It will be useful, therefore, in the next three chapters to consider firstly the types of fat in the body to which dietary fats may contribute, secondly the different types of fats in animal and plant foods and thirdly how food fats are analysed. The second part of the book, comprising Chapters 5–8, will be devoted to describing the metabolism and nutritional role of fats and their importance in health and disease.

BIBLIOGRAPHY

Brisson, G. J., *Lipids in Human Nutrition: An Appraisal of Some Dietary Concepts*, 1981, MTP, Lancaster. (A general text containing useful chapters on lipid structure, fats and oils processing and the implications of *trans* fatty acids and cholesterol in nutrition.)

Buss, D. H. and Robertson, J. (Eds), *Manual of Nutrition*, 1978, Ministry of Agriculture, Fisheries and Food, HMSO, London. (A very good reference book on the elements of food and nutrition with some useful information on food composition.)

Davidson, S., Passmore, R., Brock, J. F. and Truswell, A. S., *Human Nutrition and Dietetics*, 6th Edn, 1975, Churchill Livingstone, Edinburgh. (A good reference work for general nutrition.)

Gunstone, F. D. and Norris, F. A., *Lipids in Foods*, 1983, Pergamon Press, Oxford. (A very detailed account of lipid structures, fats and oils processing and the technological aspects of food fats.)

Gurr, M. I. and James, A. T., *Lipid Biochemistry: An Introduction*, 3rd Edn, 1980, Chapman and Hall, London. (Gives a detailed account of lipid structures and the metabolism of lipids with a final chapter on nutritional and health aspects. Useful for the nutritionist who needs to know more about the metabolic aspects of lipids.)

Chapter 2

Types of Fats in the Body and their Functions

2.1. INTRODUCTION

The types of fats in the body may be described conveniently in terms of their functions, although individual fats may have several separate roles. These functions are: structural, metabolic and storage.

As this is not a treatise on the chemical structure of fats, their chemistry will not be dealt with in detail in a separate section: rather the structures will be discussed briefly as each fat is introduced, in the context of its function. The reader who wishes to go into more detail of lipid chemistry can follow up the Bibliographical material described at the end of this chapter.

2.2. STRUCTURAL FATS

Fats play an important part in biological structures whose purpose is to provide barriers against the environment. One such barrier is the skin which is covered by a protective layer of surface lipids. These comprise tri-, di- and monoacylglycerols, wax esters, sterols, sterol esters, hydrocarbons and non-esterified fatty acids (Fig. 2.1). The importance of fats in biological barriers lies partly in their ability to exclude water. The characteristic physical feature of fats, namely their water insolubility, derives from the chemical structure of part of the fat molecule which is described as 'hydrophobic' (from the Greek, meaning 'water-hating'). In fats that are esters of fatty acids, the hydrophobic moiety is the hydrocarbon chain of the fatty acid (Figs 2.1(a) and 2.2). Some idea of the varied chemistry of the fatty acids can be obtained by looking at Fig. 2.2. The nature of the fatty acid chain plays a large part in determining the

COMPONENTS OF LIPIDS

$CH_3\ (CH_2)_x\ COOH$

(a) Fatty acid

$CH_3\ (CH_2)_y\ CH_2OH$

(b) Fatty alcohol

CH_2OH
$CHOH$
CH_2OH

(c) Glycerol

STORAGE LIPIDS (e.g. Adipose tissue, milk)

Simple glycerides

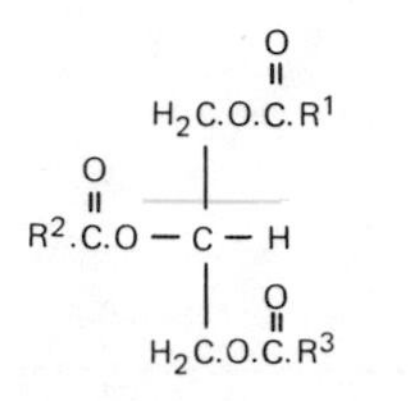

(d) Triacylglycerols

$H_2C.O.C(=O).R^1$
$R^2.C(=O).O - C - H$
$H_2C.OH$

(e) 1,2-diacylglycerols

$H_2C.OH$
$R^1.C(=O).O - C - H$
$H_2C.OH$

(f) 2-monoacylglycerols

STRUCTURAL LIPIDS (plant and animal membranes)

Phosphoglycerides *(mainly animal membranes)*

$CH_2.O.C(=O).R^1$
$R^2.C(=O).O - C - H$
$CH_2.O - P(=O)(O^-) - O - X$

(g) Phosphatidyl-x

Glycosylglycerides *(leaf chloroplast membranes)*

$H_2C.OG$
HO, O, OH, OH
$O - CH_2$
$CHOCOR^1$
CH_2OCOR^2

(h) Galactosyldiacylglycerols

Phosphosphingolipids (*animal membranes, mainly nervous tissue)*

$CH_3(CH_2)_{12}(H)C = C(H)$
$CH(OH) - CH(NH-C.O.R.) - CH_2.O - P(=O)(O^-) - O - CH_2.CH_2\overset{+}{N}(CH_3)_3$

(j) Sphingomyelin

Glycosphingolipids *(animal membranes, mainly nervous tissue)*

$CH_3(CH_2)_{12}(H)C = C(H)$
$CH(OH) - CH(NH) - CH_2 - O$ + Sugar
$NH - C = O - (CH_2)_{16 \rightarrow 22} - CH_3$

(k) Cerebroside

Steroids

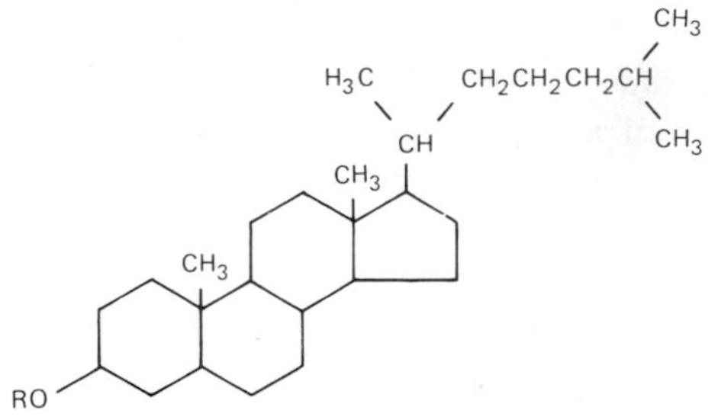

(l) Cholesterol

Sulpholipids

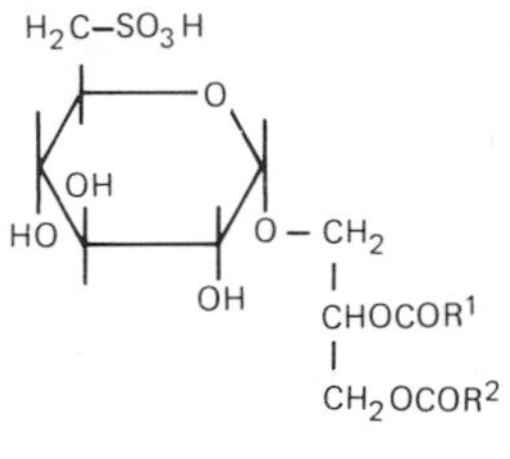

(m) Sulphoquinovosyldiacylglycerol

Fig. 2.1. Structures of some important food and body lipids. (a)–(c) *Lipid components:* Fatty acids (a) are the most widespread lipid components; they are esterified with fatty alcohols (b) to form wax esters or with glycerol (c) to form glycerides (d)–(h) and (m). (d)–(f) *Simple glycerides:* R represents the hydrocarbon chain of fatty acids; triacylglycerols (d) are the major storage lipids; di- (e) and monoacylglycerols (f) are intermediates in the synthesis or breakdown of triacylglycerols. (g) *Phosphoglycerides:* R represents the hydrocarbon chain of fatty acids; X can be one of several small molecular weight bases; the most commonly occurring is choline $(OH.CH_2.CH_2.\overset{+}{N}(CH_3)_3)$ in which case the phospholipid is phosphatidyl choline, often called by its trivial name lecithin; other bases present in naturally occurring phospholipids are ethanolamine, serine, inositol and glycerol; the positions occupied by fatty acids are numbered 1 or 2, the phosphate at position 3; when the hydroxyl group at position 2 is free and not esterified with a fatty acid, the compound is known as a lysophospholipid (e.g. lysophosphatidylcholine). (h) *Glycosylglycerides:* Here a sugar replaces the phosphate ester; R again represents fatty acid hydrocarbon chains; G may be hydrogen, in which case the lipid is called monogalactosyldiacylglycerol or another galactose molecule, in which case the lipid is digalactosyldiacylglycerol. (j) *Phosphosphingolipids:* Here the phosphorus and fatty acid are attached to the alcohol sphingosine instead of glycerol; the fatty acid is linked through an amide (nitrogen) linkage rather than the ester (oxygen) linkage of glycerolipids; the base is normally choline. (k) *Glycosphingolipids:* Instead of an ester linkage between sphingosine and phosphorus there is a glycosidic linkage between sphingosine and a number of different sugars (e.g. glucose, galactose). (l) *Cholesterol:* Cholesterol is the main steroid of animal tissues and is present as the free alcohol (in which group R is a hydrogen atom) or as cholesterol esters (in which R is a fatty acid). (m) *Sulpholipids:* These are similar in structure to galactosylglycerides (h) except that the sugar is 6-sulphoquinovose.

physical properties of the fats of which they are a part. Thus the larger the number of double bonds (higher degree of unsaturation) the lower the melting point. Within the groups of saturated fatty acids, the melting point is also lowered as the chain length decreases, or if the chain is 'branched' (Figs 2.2(d) and (e)). Table 3.10 illustrates some effects of fatty acid structure on melting point. Similar hydrocarbon chains with hydrophobic properties are seen in fatty alcohols (Fig. 2.1(b)) which are normally present as components of wax esters or in aliphatic hydrocarbons. The hydrocarbon chain is not the only hydrophobic structure found in nature. The sterol ring system is widespread and the most abundant sterol in the animal kingdom is cholesterol (Fig. 2.1(l)). Because it possesses an alcohol function, it may also form esters with fatty acids (sterol esters) and these are among the most hydrophobic of all body lipids.

The fatty acids of the skin surface lipids are unusual in two respects. Firstly, they differ in structure from almost all other body lipids in comprising significant quantities of odd-numbered carbon chains, branched chains and unsaturated acids with double bonds in unusual positions (Fig. 2.2). Secondly, a large proportion of them are present as the free (unesterified) carboxylic acids. This is unusual in living tissue, since unesterified fatty acids are toxic to cells when present in large quantities. Part of their effectiveness on the skin may lie in their ability to poison bacterial cells.

Fats also form an integral part of biological membranes. All living cells are surrounded by a membrane that provides a barrier between the cell and its environment. They also occur within the cell, providing a matrix on which much of the cell's complex chemistry takes place. In mammals, the lipids involved in membrane structures are mainly the glycerophospholipids (Fig. 2.1(g)) and unesterified ('free') cholesterol (Fig. 2.1(l)). Glycerophospholipids, like triacylglycerols (Fig. 2.1(d)) are esters of glycerol with fatty acids, but in contrast to triacylglycerols, one of the fatty acids is replaced by phosphoric acid. The parent compound is (*sn*) 1,2-diacylglycerol phosphate or phosphatidic acid (see the legend to Fig. 2.1 for an explanation of the naming and numbering of lipids). Since phosphoric acid can form diesters, a variety of different glycerophospholipids can be formed by the esterification of different 'bases' to phosphatidic acid (Fig. 2.1(g)). The most abundant, in animal tissues, is choline; thus the major mammalian phospholipid is phosphatidylcholine. The phosphocholine moiety is often called the 'polar head' group to distinguish it from the hydrophobic fatty acyl chains.

The importance of these compounds lies in their possession of chemical groupings that associate with water ('hydrophilic' groups) in juxtaposition with hydrophobic moieties. These sorts of lipids are often called 'polar' lipids or more technically 'amphiphilic' (from the Greek, meaning 'liking both') and this amphiphilic nature is of immense importance in respect of their properties in membranes and in foods. (In contrast, hydrophobic fats without polar groups such as triacylglycerols, wax esters and sterols are often called 'neutral' or 'apolar' lipids but these are imprecise terms and better avoided.)

Glycolipids are amphiphilic lipids in which the polar moiety is a sugar. They are also found in membranes, but they are less abundant in animal tissues than the phospholipids. Brain and nervous tissue are, however, particularly rich in glycolipids and phosphoglycolipids, especially those based on the alcohol sphingosine as distinct from glycerol (Figs 2.1(f) and (k)).

Current theories of biological membranes envisage that most of the lipid is present as a bimolecular sheet with the fatty acid chains in the interior of the bilayer. Membrane proteins are located at intervals at the internal or external face of the membrane or projecting through from one side to the other (Fig. 2.3). There may be polar interactions between phospholipid head groups and ionic groups on the proteins as well as hydrophobic interactions between fatty acid chains and hydrophobic amino acid sequences. Lipid molecules are quite mobile along the plane of the membrane and there may be limited movement across the membrane. Indeed, the patterns of lipid molecules on each side of some membranes are quite different, a phenomenon called 'membrane asymmetry'.

The fatty acid chains are in constant motion and the degree of molecular motion within the membrane (often referred to as 'fluidity') is influenced by the nature of the fatty acid chains, interactions between fatty acid chains and cholesterol, and interactions between proteins and lipids. As can be appreciated from Fig. 2.2, the chemistry of individual fatty acids greatly influences their shape. The presence of double bonds (Figs 2.2(f)–(n)) or of chain branching (Figs 2.2(d)–(e)) increases the bulkiness of the fatty acids and the space that they occupy. Straight chain saturated fatty acids (Figs 2.2(a)–(c)) can pack together in an almost crystalline array and molecular motion tends to be minimized compared with unsaturated or branched acids which occupy more space and are more mobile. Cholesterol plays a vital role in stabilizing hydrophobic interactions within the membrane.

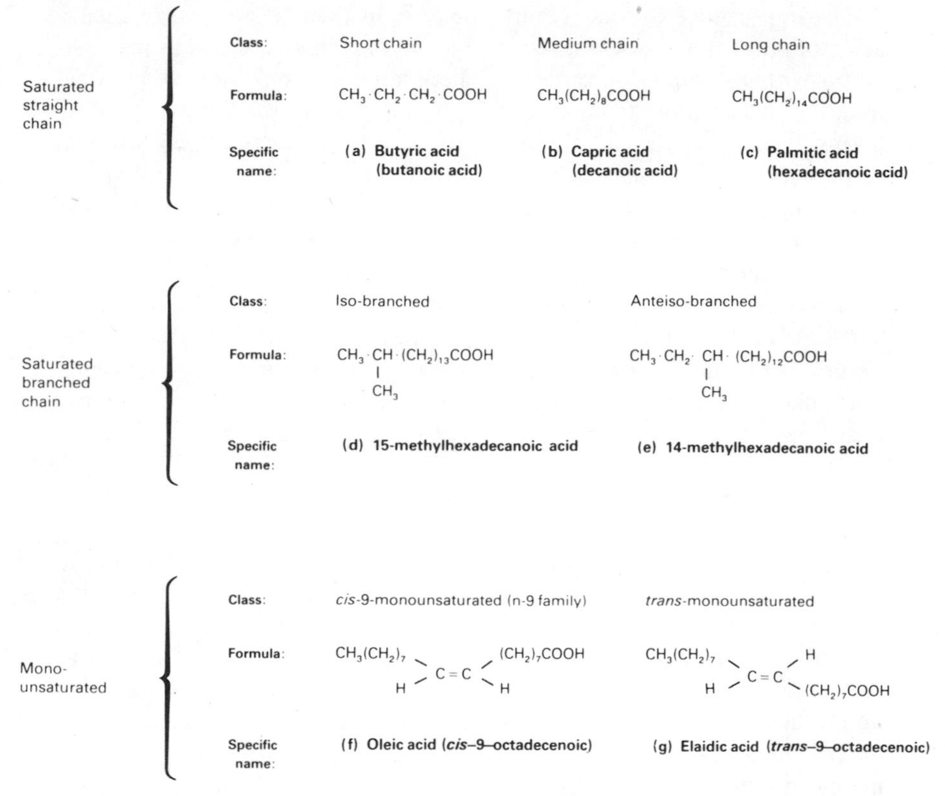

Fig. 2.2. Some fatty acids occurring in food and body lipids. The numbering of the carbon chain is always from the carboxyl group (-COOH). Thus a substituent such as a methyl group on the 4th carbon from the carboxyl group of a 16-carbon saturated acid would be 4-methyl-hexadec*an*oic acid, etc. If a double bond occurs between carbon atoms 9 and 10 of an 18 carbon acid it is called 9-octadec*en*oic acid. The nomenclature: *n*-3, *n*-6, *n*-9 describes families of fatty acid in which the last double bond in the chain is 3, 6, or 9 carbon atoms from the methyl group end of the molecule. The suffixes *c* and *t* below double bonds in the formulae denote the geometrical configuration *cis* or *trans*. The terms 'monounsaturated' and 'monoenoic' are interchangeable.

In subsequent figures and tables in this book, a common shorthand notation for fatty acids will be used. In this system, the fatty acid is identified by a number denoting the number of carbon atoms in its hydrocarbon chain, followed by a colon, followed by a number denoting the number of double bonds. Thus, stearic acid (18 carbon atoms with no double bonds) is represented as 18:0 and octadecenoic acids (18 carbon atoms with 1 double bond in an unspecified position) as 18:1, etc. If the unsaturated acid needs to be identified more precisely, it can be done by indicating the position and configuration of the double bonds,

Diunsaturated

- **Class:** *cis, cis*-9,12-diunsaturated (n-6 family) | *cis, trans*-diunsaturated
- **Formula:** $CH_3(CH_2)_4CH \underset{c}{=} CH \cdot CH_2 \cdot CH \underset{c}{=} CH \cdot (CH_2)_7COOH$ | $CH_3(CH_2)_5CH \underset{t}{=} CH \cdot CH \underset{c}{=} CH(CH_2)_7COOH$
- **Specific name:** **(h) Linoleic acid (*cis, cis*–9,12–octadecadienoic)** | **(j) *cis*–9, *trans*–11–octadecadienoic acid**

Triunsaturated

- **Class:** all-*cis*-9,12,15-triunsaturated (n-3 family)
- **Formula:** $CH_3 \cdot CH_2 \cdot CH = CH \cdot CH_2 \cdot CH = CH \cdot CH_2 \cdot CH = CH \cdot (CH_2)_7COOH$
- **Specific name:** **(k) α-linolenic acid (all *cis*–9,12,15–octadecadienoic)**

Tetraunsaturated

- **Class:** all-*cis*, 5,8,11,14-tetraunsaturated (n-6 family)
- **Formula:** $CH_3(CH_2)_4(CH = CH \cdot CH_2)_4(CH_2)_2COOH$
- **Specific name:** **(l) Arachidonic acid (all *cis*–5,8,11,14–eicosatetraenoic)**

Penta and hexaunsaturated

- **Specific names:** **(m) all–*cis*–7,10,13,16,19–docosapentaenoic acid (n–3 family)**
  (n) all–*cis*–4,7,10,13,16,19–docosahexaenoic acid (n–3 family)

thus: oleic acid, *c* 9–18:1; linoleic acid, *c* 9, *c* 12–18:2, etc. In some cases it is useful to specify the *family* to which the fatty acid belongs; thus, 18:3 may be either 18:3, *n*-3 (α-linolenic acid, *all-cis*-9, 12, 15–18:3) or 18:3, *n*-6 (γ-linolenic acid, *all-cis*-6, 9, 12–18:3).

The lipid serves to provide an insulating environment for the many metabolic activities in the membrane that involve proteins. The proteins in question may be enzymes, transporters of small molecules across the membrane or 'receptors' for substances such as hormones, antigens or nutrients. The property of 'fluidity' seems to be important in so far as it is regulated in the face of different dietary fat intakes by subtle changes in the proportions of phospholipids, cholesterol and fatty acids.

Because of the similarity of the major metabolic processes in different mammals, structural lipids do not differ enormously from one to another, although there may be minor differences in fatty acids and phospholipid polar head groups between ruminants and monogastric animals.

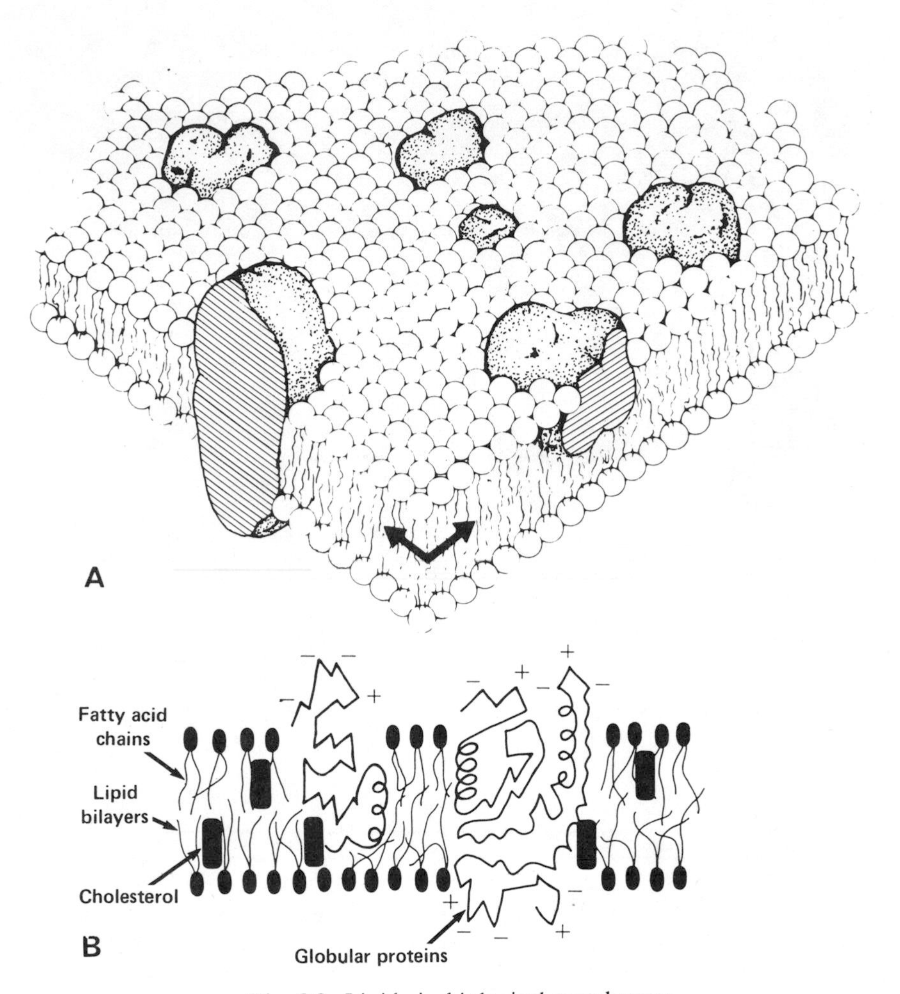

Fig. 2.3. Lipids in biological membranes.

(A) A schematic representation of what a biological membrane might look like. The small spheres are the polar head groups of phospholipid molecules which form a double layer (lipid bilayer) throughout the membrane with the fatty acid chains (represented by wavy lines) pointing inwards towards each other. Protein molecules (represented by large stippled shapes, shaded where they have been cut through in section) are inserted at intervals, through the lipid bilayer, sometimes mainly at one face or the other, sometimes extending right through.

(B) This shows the section through a membrane in diagrammatic form. In addition the scheme indicates the association of cholesterol molecules with phospholipid fatty acid chains. Protein–lipid interactions are either between polar head groups of phospholipids and hydrophilic groups of proteins or between fatty acid chains and hydrophobic regions of proteins.

(Parts A and B are reproduced from S. J. Singer, *Science*, 1972, **175**, 720–31 by kind permission of the author and the American Association for the Advancement of Science. © AAAS, 1972.)

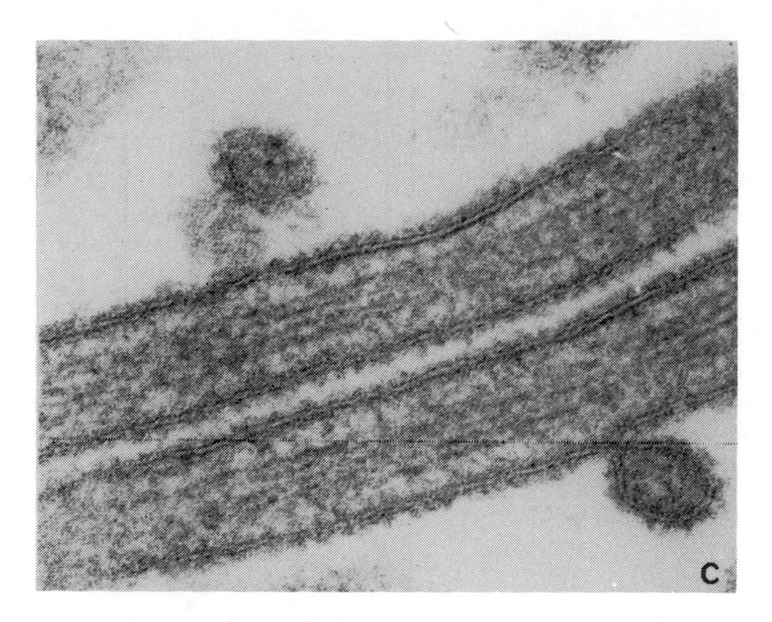

Fig. 2.3—*contd*. (C) A transmission electron micrograph of the membranes of the microvilli of the intestinal absorbing cells, illustrating clearly the dark parallel lines of the membrane bilayers.

2.3. METABOLIC FATS

The importance of lipids in membranes lies in the ability of the molecules to associate together in three-dimensional arrays. It is the physical properties of these aggregates rather than the properties of each individual molecule that are important in the overall membrane structure, although of course the chemical features of each molecule do influence the way in which the molecules interact. When we come to consider the importance of fats in metabolism, we have to consider the capacity of individual lipid molecules to undergo biochemical transformations to produce specific substances of physiological importance. Examples will be given here which will be developed in more detail in later chapters.

Fatty acids make a major contribution to the production of cellular energy (Section 6.5.2). To fulfil this function they must be released from storage fats and directed individually into metabolic pathways designed specifically to generate usable chemical energy.

Specific types of unsaturated fatty acids that are stored in membrane phospholipids can be released, and transformed into hormone-line substances called prostaglandins (Section 7.2.3.2). In this way, some 'structural' fats can be regarded as having a very special 'storage' role and the distinction between the two becomes blurred. Cholesterol is metabolized in the adrenal gland to a variety of steroid hormones and, in the liver, to a variety of bile acids that are secreted into bile and are subsequently involved in the digestion and absorption of dietary fats in the alimentary tract (Section 6.2).

Fat soluble vitamins, in so far as they participate in metabolic processes (in ways that are not yet completely defined) can also be included as 'metabolic' lipids (Section 7.3). However, they may also be stored in the liver or adipose tissue for some time before being used in these metabolic processes.

2.4. STORAGE FATS

Fatty acids in the form of simple glycerides (Figs 2.1(d)–(f)) constitute the major source of fuel in mammals. The triacylglycerols are by far the most important storage form: the partial glycerides (mono- and diacylglycerols) are for the most part intermediates in the biosynthesis or breakdown of triacylglycerols and do not accumulate in appreciable quantities in storage tissues. In the older biochemical literature and in current literature for the more general reader, you will normally find these compounds referred to as 'triglycerides'. The term 'triacylglycerol' is the officially approved nomenclature and will be used throughout this monograph. It is entirely synonymous with 'triglyceride'.

There tends to be a distinction between the types of fatty acids fulfilling a storage role (and therefore esterified in triacylglycerols) and those fulfilling a structural role (and therefore found mainly in amphiphilic lipids). Storage fats tend to contain more saturated or monounsaturated fatty acids, although this is only a general guideline since the composition of storage lipids is influenced by the fatty acid composition of the diet in monogastric animals or by the fermentative activities of the rumen microorganisms in ruminants (Section 3.2.1).

The biggest reservoir of fatty acids to supply the long-term needs of human beings for energy is the adipose tissue. Fatty acids are mobilized from this tissue to meet demands for energy at times when dietary energy is limiting. The release of stored energy is regulated by the amounts and types of different dietary components and by hormones, whose secretion may also be regulated in part by diet. Other tissues, such as the liver of mammals, can accommodate fat in the form of small globules, but only in the short term. The excessive accumulation of fat in mammalian liver is a pathological condition. However, many species of fish normally store fat in the liver or the flesh rather than in adipose tissue (Section 3.2.4).

Milk fat can also be regarded as a form of energy store, for the benefit of the new-born, and like adipose tissue fat, is composed mainly of triacylglycerols (Sections 3.2.2 and 5.3). Egg yolk lipids likewise provide a store of fuel for the developing embryo (Section 3.2.3).

Storage fat may be derived directly from the fat in the diet or it may be synthesized in the adipose tissue, mammary gland or liver from simple sugars originating from dietary carbohydrates. The capacity of these tissues to synthesize fatty acids is geared to the animal's needs and is under dietary and hormonal control. The range of fatty acids that can be made is limited. Mammalian adipose tissue normally synthesizes palmitic, stearic (saturated) and oleic (monounsaturated) acids (see Fig. 2.2). The mammary gland is unique in making short and medium chain fatty acids (chain lengths up to 12 or 14 carbon atoms), whereas all other tissues produce mainly long chain fatty acids (mainly 16 and 18 carbon atoms). The pattern of milk fatty acids is characteristic of each species. In ruminant animals the composition is markedly influenced by the activities of the rumen microorganisms (Section 3.2.2). In simple-stomached animals milk fat composition is influenced by the amount and type of fat in the diet (Section 3.2.2).

When there is little fat in the diet, the fatty acid pattern of a storage tissue is entirely dependent on its biosynthetic activity, and is characteristic of that tissue. The introduction of fat into the diet suppresses, to varying extents, the synthetic activity of the tissues, and mechanisms operate to transport dietary fatty acids into the storage fat so that the fat composition is more characteristic of the diet (Sections 3.2.2 and 6.4).

2.5. FATS IN TRANSIT

Body fats, even when they are part of the 'structural' or 'storage' pools, are in a dynamic state. There is a continuous exchange of fatty acids in membranes or in adipose tissue with fatty acids in the blood supply. Fats also have to be transferred from tissues where they are synthesised into storage or from storage pools to sites of metabolism. Fats absorbed from the alimentary tract after digestion of the diet have to be transported to sites of storage or metabolism depending on the animal's current energy needs. The route of transport is the bloodstream which is essentially an aqueous fluid in which small molecules are dissolved and macromolecules and cells are suspended. This poses a problem when fatty substances have to be accommodated. Nature has evolved methods for stabilizing fats in the bloodstream by conjugating them with proteins to form a range of transport particles called lipoproteins whose nature and metabolism will be described in rather more detail in Section 6.3.

2.6. SUMMARY

A basic understanding of the types and roles of fats in the human body is a prerequisite to the understanding of the role of fats in food and nutrition. Body fats fulfil three main functions: structural, storage and metabolic, although there is not always a complete distinction between them.

Structural fats, mainly phospholipids, glycolipids and cholesterol, form an integral part of biological membranes, which act as barriers between one body compartment and another and also are the sites of production of many biochemical substances important in metabolism. Storage fats are mainly triacylglycerols and provide a long-term supply of energy in time of food shortage. The triacylglycerols are stored in specialized cells, the fat cells, in adipose tissue. The accumulation of fats in, and their release from, the adipose tissue are controlled by an interplay of dietary and hormonal factors. The fat molecules in membranes and in the fat depots are present in large aggregates. Individual fat molecules may be released from either of these sources by specific enzymic processes and converted into metabolites with important functions. Examples are the prostaglandins from essential fatty acids, bile acids and steroid hormones from cholesterol.

Even though fats are insoluble in water, they need to be transported between tissues in the body in the bloodstream, which is essentially an aqueous medium. This is effected by binding the fats to proteins in the form of lipoproteins.

BIBLIOGRAPHY

Ansell, G. B., Dawson, R. M. C. and Hawthorne, J. N., *Form and Function of Phospholipids*, 1973, Elsevier, Amsterdam. (A very thorough account of phospholipid structures, occurrence and metabolism.)

Brisson, G. J., *Lipids in Human Nutrition: An Appraisal of Some Dietary Concepts*, 1981, MTP, Lancaster. (Chapter 1 gives a useful account of the structures and properties of the various fats found in the body and in food.)

Gurr, M. I., The nutritional significance of lipids. In: *Developments in Dairy Chemistry – 2*, ed. P. F. Fox, 1983, Applied Science, London, Chapter 8. (Contains a condensed account of the contents of this chapter and an extensive reference list for further reading.)

Gurr, M. I. and James, A. T., *Lipid Biochemistry: An Introduction*, 3rd Edn, 1980, Chapman and Hall, London. (Chapter 7 gives a brief account of the role of lipids in food and some references for further reading.)

Chapter 3

Fats in Food

3.1. SOURCES OF DIETARY FATS

Omnivorous man derives his dietary fats from plant and animal tissues. Just as these contain 'structural' and 'storage' fats, so can man's dietary fats be considered in these terms. Food in general contains two types of fats. 'Visible' fats derive from the adipose tissue and the milk of animals or from seed oils that are in effect the storage lipids of the plant. Fabricated fats, like margarines, are themselves derived from plant or animal storage fats which have been modified by processing. 'Hidden' fats derive from the membranes of animal or plant tissues. It should also be remembered that a great deal of fat in prepared food is hidden in the sense that it is incorporated during cooking, e.g. in cakes, biscuits, potato crisps or in the formulation of processed meats and sausages. This fat will be largely storage triacylglycerol, not structural fat.

3.2. COMPOSITION OF FOOD FATS

3.2.1. Adipose Tissue

The fat in adipose tissue is almost entirely composed of triacylglycerols but the tissue also stores smaller amounts of cholesterol, fat-soluble vitamins A, D, E and K and also fat-soluble substances that may enter the food from the environment such as pesticides and antioxidants added to food.

The composition of the adipose tissue of monogastric (simple-stomached) animals, of which pigs and poultry are economically the most important, is markedly affected by diet. When these animals are fed low-fat cereal based diets, as in traditional farming practice, the adipose tissue fatty acids are synthesized in the body from dietary carbohydrates and the depot* fat is composed mainly of saturated and monounsaturated fatty acids (see Table 3.1). However, cereals contain structural lipids that influence depot fat composition to some extent, depending on the type of cereal. Inclusion of vegetable oils, such as soybean oil, in the diet results in higher proportions of linoleic acid and lower proportions of oleic and palmitic acids than the inclusion of tallow which tends to give a depot fat similar in composition to that of animals fed cereals. Feeding unsaturated fat supplements or including appreciable amounts of copper in pig diets tends to result in soft backfat.

The adipose tissue of ruminants is less variable than that of simple-stomached animals because about 9/10 of the dietary unsaturated fatty acids are hydrogenated (i.e. converted into relatively more saturated fatty acids) by microorganisms in the rumen. It contains a larger proportion of saturated and monounsaturated fatty acids and a lower proportion of polyunsaturated fatty acids than the adipose tissue of monogastric animals. It also contains more *trans* unsaturated and branched chain fatty acids. *Trans* double bonds are formed by rearrangement of the more abundant *cis* double bonds during the process of hydrogenation by enzymes in certain types of rumen microorganisms. The biochemical process is illustrated in Fig. 3.1. Similar chemical changes occur when unsaturated fats are hydrogenated chemically in industrial processes as described in Section 3.3. The amounts and types of these more unusual fatty acids that are present in ruminant products depends on the species of animal and the way in which its diet affects the activities of the rumen microflora. For example, feeding sheep a diet rich in barley leads to the production of much higher proportions of branched chain fatty acids in their storage fat than when they are grazing pasture, and a practical consequence is an increased softness of the depot fat.

In traditional agricultural practice, ruminants in any case are fed diets relatively low in fat, comprising not more than about 2–5% dietary digestible energy.† Under these circumstances, most of the body fats (except those which are dietary essentials — see Chapter 7) are synthesized within the body from the products of microbial fermentation of

* The terms 'adipose tissue' fat and 'depot' fat are used interchangeably here.

† In this book, the word 'energy' is used in the nutritionist's sense of the chemical energy of the diet which can supply the nutritional needs of the animal for fuelling its metabolic processes. It is not used in the layman's sense of 'activity' or 'liveliness'.

TABLE 3.1
The Fatty Acid Composition of Some Animal Storage Fats used in Human Foods Showing the Influence of Different Feeding Practices (g 100 g^{-1} Total Fatty Acids)

	Pig (lard)			*Poultry*		*Beef (suet)*		*Lamb*		*Cod liver oil*
Fatty acid	*A*	*B*	*C*	*D*	*E*	*F*	*G*	*H*	*I*	*J*
Myristic (14:0)	1	1	1	1	1	3	3	3	4	6
Palmitic (16:0)	29	21	21	27	22	26	20	21	19	13
Palmitoleic (16:1)	3	3	4	9	5	9	4	4	6	13
Stearic (18:0)	15	12	17	7	6	8	10	20	16	3
Oleic (18:1)	43	46	54	45	27	45	33	41	37	20
Linoleic (18:2)	9	16	3	11	35	2	23	5	12	2
Long chain, monounsaturated (20:1, 22:1)	0	0	0	0	0	0	0	0	0	18
Long chain polyunsaturated (20:5, 22:5, 22:6)	0	0	0	0	0	0	0	0	0	20
Others	0	1	0	0	4	7	7	6	6	5

Adapted from M. I. Gurr, Agricultural aspects of lipids, In: *The Lipid Handbook*, 1984, Chapman and Hall, London and M. I. Gurr, The nutritional significance of lipids, In: *Developments in Dairy Chemistry — 2: Lipids*, ed. P. F. Fox, 1983, Applied Science Publishers, London.

A, Pig fed low fat cereal based diet: B, Pig fed high fat diet containing soybean oil; C, pig fed high fat diet containing beef tallow; D, poultry fed low fat cereal based diet; E, poultry fed high fat diet containing soybean oil; F, cattle fed diet based on hay; G, cattle fed diet containing 'protected' safflower oil; H, lambs fed cereal based concentrate diet; I, lambs fed diet containing 'protected' safflower oil. J, cod liver oil.

Note the increases in the proportion of linoleic acid in pigs and poultry fed polyunsaturated vegetable oils and cattle and lambs fed 'protected' vegetable oils. Also, note the low proportion of linoleic acid in the fish oil and the correspondingly high proportion of long chain polyunsaturated fatty acid.

dietary carbohydrate in the rumen. A more recent trend, to improve the efficiency of animal production, is to feed ruminants, especially dairy cows, fat supplements or whole oil seeds. In this way modest changes in depot fat composition can be affected, but more extensive changes are brought about by feeding so-called 'protected' fat. In this process, fat particles are coated with protein which is crosslinked by treatment with formaldehyde. This protects the fat against degradation in the rumen but

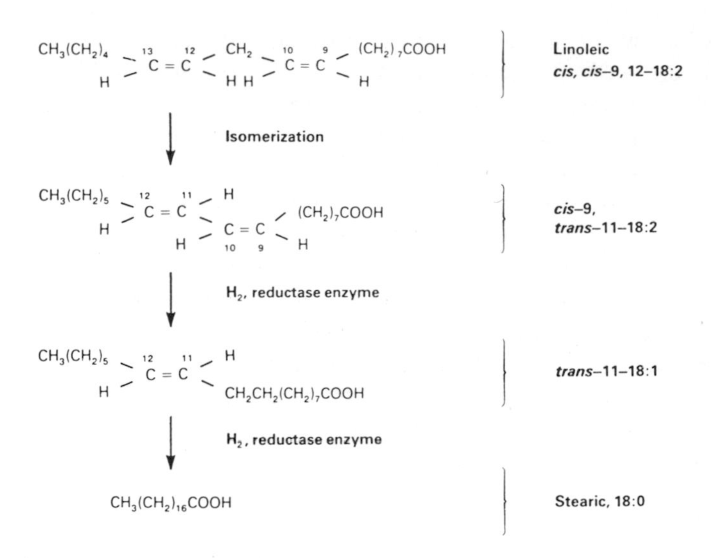

Fig. 3.1. Steps in the biohydrogenation of dietary fatty acids in the rumen of ruminant animals. (Adapted from M. I. Gurr and A. T. James, *Lipid Biochemistry: An Introduction*, 1980, Chapman and Hall, London.)

on reaching the acidic environment of the true stomach (abomasum), the protein is digested and the fat can pass unchanged into the small intestine. There it can be digested and absorbed by mechanisms identical to those in simple-stomached animals and which are described in a little more detail in Chapter 6. In this way, large increases in, for example, the proportion of linoleic acid can be brought about in ruminant depot fats, milk and to a lesser extent in the carcass fats (Table 3.1).

Research into animal nutrition has led to changing farming practice, resulting in increased food conversion efficiency and faster growth in cattle, pigs and poultry. Consumers, however, are increasingly demanding leaner carcasses (the fat thickness at the last rib in pigs has fallen by 0·5 mm each year for the last 10 years). In using fats in feeds, therefore, a compromise has to be reached between the advantages of increased efficiency and faster growth and the tendency to lay down excessive carcass fat. In the pig industry, the trend is towards the increasing use of boars, which are leaner at corresponding weights than castrates.

A consequence of the slower growth that yields a leaner carcass is a relatively more unsaturated fatty acid composition in adipose tissue (in particular, a higher proportion of linoleic acid) and therefore a softer carcass fat, which is not desirable to the average consumer. The same effect is produced by feeding unsaturated fat supplements. This may be overcome to some extent, while still feeding energy dense diets, by incorporating various more-saturated fats, such as hydrogenated fish oil or hydrogenated tallow into animal diets.

As well as producing fat of undesired appearance and texture, an overabundance of unsaturated fat limits the storage life of meat by oxidative breakdown of unsaturated fatty acids, leading to the production of peroxides (see Section 3.3.7) and eventually to rancidity. The oxidative instability and taste problems associated with products from ruminants fed protected oilseeds is a major factor contributing to the failure of such products to become established in the market place. Even when the consumer requires leaner carcasses or modified fat products, the associated changes in colour and softness of the fat may be unacceptable.

3.2.2. Milk

Cow's milk and dairy products derived from it are important in human foods and in the UK, 1/3 of dietary fat intake is from milk and dairy products. The only milk of quantitative importance in the adult human diet in the UK and in most Western Countries is cow's milk. In some countries, goat's milk and sheep's milk are of considerable importance and there is increasing interest in these milks in the UK.

Milk fat is composed mainly of triacylglycerols which are present in the milk as an emulsion in which the fat globules are stabilized by a surrounding membrane composed of proteins and phospholipids. The fat globules also contain smaller amounts of cholesterol and fat soluble vitamins, mainly vitamins A and D (Table 3.2). Although there is a large range in size of the milk fat globules, goat's milk possesses on average smaller fat globules than cow's milk: It has been inferred that this enables the fat to be more efficiently digested by man but this possibility does not seem to have been subjected to rigorous scientific proof.

The fatty acid compositions of cow's and goat's milk are characterized by a high proportion of short and medium chain fatty acids, long chain saturated and monounsaturated acids and very small proportions of polyunsaturated fatty acids (Table 3.3). Ruminant milks also contain small quantities of a wide variety of branched and odd chain fatty acids depending on the feeding system (see Section 3.2.1).

Milk yields, which have been steadily rising over the years, depend on breed, stage of lactation and on the management, particularly the diet, of the cattle. Strategies to increase yield often result in lower milk fat

TABLE 3.2
The Lipid Composition of the Fat Globule and the Fat Globule Membrane of Human and Cow's Milk (g 100 g^{-1} Total Lipids)

Lipid	*Human*		*Bovine*	
	Fat globule	*Fat globule membrane*	*Fat globule*	*Fat globule membrane*
Hydrocarbons	trace	trace	trace	1·2
Sterol esters	trace	trace	trace	0·1–0·8
Sterols	0·2	0·7	0·2–0·4	0·2–5·2
Triacylglycerols	98·1	58·2	97–98	
Diacylglycerols	0·7	8·1	0·3–0·6	53–74
Monoacylglycerols	trace	0·6	trace	
Free fatty acids	0·4	7.3	0·1–0·4	0·6–6·3
Phospholipids	0·3	23·4	0·2–1·0	12–45

Adapted from M. I. Gurr, Review of the progress of dairy science: human and artificial milks for infant feeding, *J. Dairy Research*, 1981, **48,** 519–54.

content and altered fat composition. The biggest changes occur when cows are offered highly digestible diets containing large proportions of cereal grains, high quality forages or fat supplements in order to increase the intake of digestible energy. The effect of increasing the concentration of digestible carbohydrate is almost always to decrease milk fat percentage, although not necessarily milk fat yield, because of higher overall milk yield. The effects of fat supplements are more variable. Feeding unprotected fat often decreases milk fat content because of inhibition by dietary fatty acids of rumen microorganisms, resulting in decreased fibre digestibility. Feeding protected fats often overcomes these problems. Higher amounts of fat can be fed and an increase in milk fat content is achieved. Feeding fat to ruminants depresses the concentrations of short and medium chain fatty acids in milk by inhibiting the biosynthesis of fatty acids within the mammary gland and allowing the incorporation of dietary long chain fatty acids into milk. If the protected fat contains linoleic acid, the concentration of this acid in milk can be increased.

Butter is a common food fat containing 15% water as an emulsion in oil. The fat is derived entirely from cow's milk and its composition does not normally vary greatly, in contrast to the composition of margarine, discussed in Section 3.2.8. However, the fatty acid composition of butter is susceptible to modification if the fat in the cow's diet is protected from hydrogenation in the rumen as described earlier.

TABLE 3.3
Fatty Acid Composition (g 100 g^{-1} Total Fatty Acids) of Some Milks of Importance in Human Nutrition

Fatty acid[b]		*Cow*[a]		*Goat*	*Sheep*	*Human*
		A	*B*			
Saturated	4:0	3	10 (4:0–12:0)	2	4	0
	6:0	2		2	3	0
	8:0	1		3	3	0
	10:0	3		9	9	1
	12:0	4		5	5	5
	14:0	12	8	11	12	7
	16:0	26	17	27	25	27
	18:0	11	8	10	9	10
		62	43	69	70	50
Monounsaturated	14:1	1	0	1	1	1
	16:1	3	0	2	3	4
	18:1	28	21	26	20	35
		32	21	29	24	40
Polyunsaturated	18:2	2[e]	35[f]	2	2	7
	18:3[c]	1	1	0	1	1
	20:4	0	0	0	0	trace
		3	36	2	3	8
Others[d]		3	0	0	3	2

Adapted from M. I. Gurr, Review of the progress of dairy science: human and artificial milks for infant feeding. *J. Dairy Research*, 1981, **48,** 519–54.

[a] A, cow grazing; B, cow fed supplement of protected safflower oil.
[b] For shorthand system for naming fatty acids see caption to Fig. 2.2.
[c] Mainly α-linolenic acid (9, 12, 15–18:3); human milk may contain a trace of γ-linolenic acid (6, 9, 12–18:3).
[d] Includes odd-chain and branched chain fatty acids in the case of ruminant milks and very long chain polyunsaturated fatty acids (22:4, *n*-6; 22:5, *n*-6; 20:5, *n*-3; 22:5, *n*-3; 22:6, *n*-3) in the case of human milk.
[e] Contains linoleic acid (*c*, *c*-9, 12-18:2) and other positional and geometric 18:2 isomers with no EFA activity.
[f] Mainly linoleic acid (*c*, *c*-9, 12-18:2).

Cream is another example of an emulsion of milk fat in water whose textural properties and quantitative contribution of fat to the diet depend on the total concentration of fat in the emulsion. Fat emulsions markedly influence the palatability of food as discussed in Section 5.2.1.

3.2.3. Eggs

The egg of the domestic fowl provides a significant source of fat in many human diets. The yolk is rich in fats whose biological function is to sustain and nourish the developing embryo. The lipid composition is summarized in Table 3.4 and these lipids are present in the form of both low density and high density lipoproteins. The latter are called lipovitellins. As Table 3.4 shows, the fatty acid composition of the glycerolipids is predominantly saturated and monounsaturated, although the phospholipid fraction contributes polyunsaturated fatty acids to give about 10% of the total. Cholesterol is a major component and one egg on average provides about 300 mg cholesterol.

The importance of eggs in food lies not only in the nutritive value of the fats but in the contribution made by the liproprotein to the structure of food, particularly to the textural quality of cake after baking.

3.2.4. Fish Oils

Fish can be classified broadly into 'lean' fish that store their reserve fats as triacylglycerols in the liver (e.g. cod) and 'fatty' fish whose triacylglycerols are stored in the flesh (e.g. mackerel, herring). They have no adipose tissue. Some fish (e.g. the Orange Roughy, found in the Pacific) store their reserve oil as wax esters which have poor nutritive value.

Although fish oils may differ in composition between species, and according to diet, they have some common characteristics that distinguish them from mammalian and avian fats. They have a high content of fatty acids with 20 or more carbon atoms and are rich in polyunsaturated fatty acids of the *n*-3 family (see Figs 2.2(m)–(n), Table 3.1). During partial hydrogenation, a process necessary when fish oils are used in the food industry to achieve the desired physical properties and keeping quality, the latter are largely converted into saturated and isomeric (positional and geometric) monounsaturated fatty acids. The diet of fish is the main factor determining the fatty acid composition of their body fats and may itself be influenced by factors such as water temperature and whether the fish are free-living or farmed.

TABLE 3.4(A)
Lipid Composition of Egg Yolk Lipoproteins (g 100 g^{-1})

	Protein	*Phospholipid*	*Cholesterol*	*Cholesterol ester*	*Triacylglycerol*
High density lipoprotein (β-lipovitellin)	78	12	0·9	0·1	9
Low density lipoprotein	18	22	1·8	0·2	58

TABLE 3.4(B)
Fatty Acid Composition of Egg Yolk Lipids (g 100 g^{-1} Total Fatty Acids)

Fatty acid	*Whole egg*	*Low density lipoprotein*
16:0	29	32
16:1	4	8
18:0	9	8
18:1	43	45
18:2	11	7
Others	4	—

Adapted from A. R. Johnson and J. B. Davenport, *Biochemistry and Methodology of Lipids*, 1971, John Wiley & Sons, New York.

3.2.5. Muscle and other Animal Structural Fats

Fats eaten in muscle meats comprise mainly phospholipids and cholesterol, although in many meat animals the muscles are infiltrated with fats (marbling) which are mainly triacylglycerols and which have a fatty composition similar to the storage fat. The fatty acid composition of the lean meat is less variable and less susceptible to dietary influence than that of the storage fats. There is a high proportion of polyunsaturated fatty acids, even in ruminant muscle and in particular, muscle fats are a major source of dietary arachidonic acid (Table 3.5). Despite the homeostatic mechanisms tending to keep the composition of biological membranes within certain bounds consistent with optimal membrane function, the composition of these fats can be manipulated to a small degree by feeding monogastric animals diets containing different fat supplements and ruminants protected fats.

TABLE 3.5
The Fatty Acid Composition of Some Structural Lipids Occurring in Human Foods ($g\ 100\ g^{-1}$ Total Fatty Acids)

	Beef		*Lamb*		*Chicken*		*Pork*	*Cod*	*Green leaves*
Fatty acid	*A*	*B*	*C*	*D*	*E*	*F*	*G*	*H*	*J*
Palmitic (16:0)	16	14	22	22	23	25	19	22	13
Palmitoleic (16:1)	2	2	2	1	6	3	2	2	3
Stearic (18:0)	11	14	13	18	12	17	12	4	trace
Oleic (18:1)	20	5	30	28	33	26	19	11	7
Linoleic (18:2)	26	47	18	1	18	15	26	1	16
α-Linolenic (18:3)	1	1	4	0	1	1	0	trace	56
Arachidonic (20:4)	13	11	7	4	6	6	8	4	0
Long chain, polyunsaturated (20:5, 22:5, 22:6)	0	0	0	14	0	6	0	52	0
Others	11	6	4	12	1	1	14	4	5

Adapted from M. I. Gurr, Agricultural aspects of lipids, in: *The Lipid Handbook*, 1984, Chapman and Hall, London and M. I. Gurr, The nutritional significance of lipids, in: *Developments in Dairy Chemistry — 2: Lipids*, ed. P. F. Fox, 1983, Applied Science Publishers, London.

A, Muscle, cattle fed low fat diet; B, muscle, cattle fed 'protected' safflower oil diet; C, muscle; D, brain; E, muscle; F, liver; G, muscle; H, flesh.

Note the low content of linoleic acid and the high content of long chain polyunsaturated acids in brain and fish flesh and the high content of α-linolenic acid in green leaves.

Liver fatty acids, though mainly present in membrane phospholipids, are influenced by the presence of variable amounts of triacylglycerols in the organ. Therefore, the fatty acid composition of liver tends to be intermediate between that of the storage and the muscle fats. Liver is also a relatively rich source of dietary cholesterol. Brain, though not an important form of meat in the UK diet, contains a high concentration (8%) of structural fat, with a fatty acid composition somewhat different from that of other structural fats.

3.2.6. Seed Oils

The seed oil is a form of energy storage for many plants, just as the adipose tissue is for animals. Most oil-bearing plants store their fat in the form of triacylglycerols although some, like jojoba, store wax-esters. The latter are poorly digested and have little nutritive value. Triacylglycerols are stored mainly in the fleshy fruit exocarp (e.g. avocado) or in the seed endosperm (e.g. rape). Some, like palm, store triacylglycerols in both the exocarp (palm oil) and the endosperm (palm kernel oil). The oil is present in droplets in the cytoplasm of the seed cells. These droplets are known as oil bodies and are surrounded by a membrane composed of phospholipids and proteins, not unlike the milk fat globules.

Seed oils vary widely in fatty acid composition. One fatty acid often predominates, is frequently of unusual structure, and is characteristic of a particular plant family (Table 3.6).

The commercially important seed oils (Table 3.7) are generally, however, those in which the predominant fatty acids are the common ones: palmitic, stearic, oleic and linoleic acids. Exceptions are coconut and palm kernel oils, which are unusual in containing saturated medium chain length fatty acids. Elsewhere in nature, only milk contains these fatty acids. The characteristic seed oil fatty acids are esterified in specific positions on the triacylglycerol molecule and do not occur in the structural lipids. In plants that store triacylglycerols in both exocarp and endosperm, there is usually a marked difference in the fatty acid composition of triacylglycerols from the two locations (Table 3.6).

Seed oils also contain variable amounts of phospholipids, chlorophylls, carotenoids, tocopherols and plant sterols such as β-sitosterol, although the latter is not absorbed from the human gut. In addition, some may contain unusual fatty acids which if ingested in large amounts, may have the toxic or otherwise undesirable metabolic effects described in Section 8.8.

Total world production of seed oils in 1981/82 was 43 million tonnes.

TABLE 3.6
The Fatty Acid Composition ($g\,100\,g^{-1}$ Total Fatty Acids) of Some Vegetable Oils Used in Human Foods

	Coconut	*Corn*	*Olive*	*Palm*	*Palm kernel*	*Peanut*	*Rape*		*Soybean*	*Sunflower*
							Low erucic	*High erucic*		
8:0	8	0	0	0	4	0	0	0	0	0
10:0	7	0	0	0	4	0	0	0	0	0
12:0	48	0	0	trace	45	trace	0	0	trace	trace
14:0	16	1	trace	1	18	1	trace	trace	trace	trace
16:0	9	14	12	42	9	11	4	4	10	6
18:0	2	2	2	4	3	3	1	1	4	6
20:0	1	trace	trace	trace	0	1	1	1	trace	trace
22:0	0	trace	0	0	0	3	trace	trace	trace	trace
16:1	trace	trace	1	trace	0	trace	2	trace	trace	trace
18:1	7	30	72	43	15	49	54	24	25	33
18:2	2	50	11	8	2	29	23	16	52	52
18:3	0	2	1	trace	0	1	10	11	7	trace
Others	0	1	1	2	0	2	5	10(20:1) 33(22:1)	2	3

Adapted from M. I. Gurr, Agricultural aspects of lipids. In: *The Lipid Handbook*, 1984, Chapman and Hall, London, chapter 8, by kind permission of the publishers.

TABLE 3.7
Some Seed Crops Important in Human Foods

Seed	*Oil content, %*	*Major fatty acids*	*World total of oil production in 1982/83 (tonnes)*	*Chief producing areas*	*Major food uses*
Soybean	13–20	18:2	14,700,000	USA, Brazil, China	Margarine, cooking oil, salad oil, ice cream
Groundnut (peanut)	45	18:2	2,900,000	India, China, Africa, USA	Margarine, cooking oil, salad oil, ice cream
Coconut	63	12:0	3,200,000	Philippines, Indonesia	Margarine, cooking oil, salad oil
Oil palm:					
Palm oil	50[a]	16:0, 18:1	5,000,000	West Africa, Malaysia, Indonesia	Margarine, shortenings, biscuit fats, frying fat, confectionary fats, ice cream
Palm kernel oil	50[b]	12:0	900,000		
Rape	35–40	22:1 18:1 in zero erucic varieties	4,600,000	India, China, Canada, Poland, France, Sweden, UK	Margarine, cooking oils, salad oils
Cotton	15–23	18:2, 18:1	3,300,000	USSR, China, UK	Margarine, cooking oils
Olive	15	18:1	2,000,000	Italy, Spain, Greece	Salad oils, preserving oils
Sesame	50	18:1, 18:2	660,000	India, China, Mexico	Table oils

Adapted from M. I. Gurr, The biosynthesis of triacylglycerols. In: *The Biochemistry of Plants*, vol. 4, eds P. K. Stumpf and E. Conn, 1980, Academic Press, New York, pp. 205–48.
[a] Percentage of mesocarp. [b] Percentage of kernel.

Of the several hundred varieties of plants known to have oil bearing seeds, only 12 are important commercially and of these, three are used for industrial purposes other than as edible oils.

Demands for food fats with particular fatty acid composition or at a lower price could stimulate the agricultural production of oil seed crops that have not so far been commercially exploited (e.g. lupin or evening primrose) or the production of well known crops in countries that have not previously grown them (e.g. sunflowers in southern England). This theme will be further developed in a later chapter.

3.2.7. Plant Leaves

The leaves of higher plants contain up to 7% of their dry weight as fats, some of which are present as surface lipids, the others as components of leaf cells, especially the chloroplast membranes.

While all plant membranes contain phospholipids, the characteristic and most abundant membrane lipids in green tissue are the glycolipids in which the predominant sugar is galactose (Fig. 2.1(h)). Mono- and digalactosyldiacylglycerols together contribute about half the lipids in chloroplast membranes, chlorophyll about 1/5 and sulphoquinovosyl diacylglycerol (the 'plant sulpholipid') about 1/20. Because of the wide distribution of green plants across the surface of the earth, these plant lipids are among the world's most abundant organic compounds. Other minor lipids present in leaves are some plant sterols, acylated sterol glycosides and the carotenoids, many of which are nutritionally important as precursors of vitamin A.

The fatty acid composition of plant membrane lipids is very simple and varies little between different types of leaves. Six fatty acids generally account for over 90% of the total: palmitic, hexadecenoic, oleic, linoleic and α-linolenic (Table 3.5). Of these, α-linolenic is quantitatively the most important. Hexadecenoic acid is a mixture of isomers and contains an unusual isomer with a *trans* double bond between carbon atoms 3 and 4. This fatty acid is found exclusively as a component of the phospholipid phosphatidyl glycerol. The surface lipids contain a wider spectrum of fatty acids with chain lengths between C_{10} and C_{30} present in wax-esters and as non-esterified fatty acids, while the cutins contain a high proportion of C_{18} hydroxyacids.

3.2.8. Manufactured Fats

Margarine was one of the earlier attempts to simulate a natural product, butter. Originally invented by the Frenchman, Mège Mouriès in 1859, it was made by melting down various animal fats and churning them with cream. Modern margarine, like butter, is an emulsion of water and fat. The aqueous component is skimmed milk, slightly soured by microorganisms to enhance the flavour. The fat component is a blend of fats and oils from both vegetable and animal sources. The most commonly used vegetable fats are palm, soybean, groundnut, coconut, sunflower, cottonseed and rapeseed, while the animal fats are mainly beef and mutton tallow, herring and pilchard oils. All margarines are made from a blend of several of these, the blend varying according to the supply and cost of oils at a given time. Therefore, even within a single brand, the fat blend may change from batch to batch and the fatty acid composition will vary between certain limits designed to maintain the physical properties of that brand. In addition, manufacturing processes designed to modify the physical properties of the oils so as to achieve the desired properties of the product will alter the fatty acid composition of oils in the blend. These include catalytic hydrogenation (which results in a decrease in the number of double bonds, an increase in the proportion of *trans* double bonds and a randomization of double bond positions along the chain), interesterification (which randomizes the positions of the fatty acids on the triacylglycerol molecules) and fractionation (which separates out fats of different melting points). The fatty acid compositions of different margarines listed in Table 3.8 should therefore be regarded as examples of the fatty acid patterns to be found in margarines of different types and not to be regarded as absolute.

Margarines are not the only examples of manufactured fats. With the skilful use of modern emulsifier technology, the aqueous phase can be increased from about 15% to about 50% to give a low fat spread that may have applications in energy reduced diets (see Sections 5.2.2 and 8.3.3). One low fat spread on the market in Sweden employs a mixture of dairy and vegetable fats to give a product relatively rich in polyunsaturated fatty acids yet with a taste more closely akin to dairy products. This is a more economical way of modifying the fatty acid composition than producing butter from cows fed protected fat as described in Sections 3.2.1. and 3.2.2. (Table 3.8).

In contrast to margarines and low fat spreads, 'shortening' fats do not contain an aqueous phase. The term includes all commercial fats and oils except oil-derived products such as margarines and other high fat products containing non-fat materials. Their role is to 'shorten' or tenderize baked foods by preventing the cohesion of wheat gluten strands during mixing. They are normally made from one or more partially

TABLE 3.8
The Approximate Fatty Acid Composition of Some Spreading Fats (g 100 g^{-1} Total Fatty Acids)

	Butter		*Hard margarine*		*Soft margarine*		*Polyunsaturated margarine*	*Blended low fat spread*
Fatty acid	*A*	*B*	*C*	*D*	*E*	*F*	*G*	*H*
4:0–12:0	13	6	trace	trace	1	2	3	12
14:0	12	8	6	1	5	1	1	9
16:0	26	17	20	28	16	24	11	24
18:0	11	8	8	7	5	5	9	9
20:0	0	0	2	1	2	1	1	0
22:0	0	0	2	1	3	1	1	0
16:1	} 30	21	6	1	6	1	trace	} 23
18:1			22	42	25	37	18	
20:1	0	0	9	2	7	1	1	0
22:1	0	0	9	4	9	4	1	0
18:2	2	35	5	10	9	21	53	14
18:3	trace	trace	trace	1	trace	2	1	trace
20:4+20:5	0	0	7	1	7	trace	trace	0
22:5+22:6	0	0	4	1	5	trace	trace	0
Others	4	4	0	0	0	0	0	9

Compiled from data supplied by colleagues, from the author's own laboratory and from M. I. Gurr and A. T. James, *Lipid Biochemistry: An Introduction*, 1980, Chapman and Hall, London.

A, Butter made from milk produced by grazing cows; B, butter made from milk produced by cows fed 'protected' vegetable oil supplements (note the increased proportion of linoleic acid); C, E, manufactured from a blend of animal and vegetable oils; D, F, manufactured from a blend of vegetable oils only; G, a well known brand made from sunflower seed oil and a small amount of hydrogenated vegetable oil hardstock; H, the fat phase is about 50% of the total solids and is composed from a blend of dairy fat and a polyunsaturated vegetable oil.

hydrogenated vegetable oil base stocks or mixtures of such stocks with animal fat base stocks. They are to be found in baked goods such as biscuits, cakes, doughnuts, Danish and puff pastries, breads and rolls, icings, imitation creams, fried and frozen foods.

3.3. HOW INDUSTRIAL AND DOMESTIC PROCESSES CAN MODIFY THE COMPOSITION OF FOOD FATS

Many foods are inedible unless processed in some way. Although 'processing' is often associated by the layman with large scale industrial handling with complex machinery, the term strictly refers to any manipulation that alters the food in any way and this includes domestic preparation and cooking. The reasons for processing are many and various. Fats in the human diet may be consumed as an integral part of the food, for example in a piece of streaky bacon or a slice of avocado pear or in a more refined form either as a spread such as butter, as an aid to cooking, such as a frying oil or as the raw material for fabricated products such as shortenings and margarines. The more refined the form in which the fat is required, the larger is the number of processing steps likely to be. They may range from being relatively simple and mild to rather harsh treatments in some cases.

There is no intention here to give a detailed account of oils and fats processing. A good account is provided by Gunstone & Norris (see Bibliography). Each process will be described briefly to give some understanding of the potential for modifying fat composition and therefore possibly influencing nutritive value.

3.3.1. Extraction, Separation and Purification

Plant oils, as we have seen in Section 3.2.6, are locked up in the structure of the seed and the oil itself can only be obtained by pressing the seeds or by solvent extraction after first heating, dehulling and chopping into flakes. In the solvent extraction hydrocarbons such as hexane are used that are sufficiently volatile to be removed completely.

The main objective is to obtain a 'clean' product which is composed mainly of triacylglycerols. Many constituents of the crude oil extract are undesirable and the oil is subjected to an extensive refining process. 'Degumming' removes a variety of substances such as phospholipids and polysaccharides that would separate out from the oil upon storage. They, as well as chlorophylls or minerals, are removed into aqueous solution by heating at 60–90°C with phosphoric acid and water followed by centrifugation. This is followed by an alkaline treatment ('alkali refining') to remove non-esterified fatty acids as their sodium soaps. This is necessary because during extraction of the oil, enzymes called 'lipases' that are naturally present in the plant tissue become activated and cause the release of fatty acids from the plant lipids (Fig. 3.2). Indeed to observe lipolytic breakdown, it is not always necessary even to extract the tissue: almost the first chemical change to occur when vegetables are picked is the release of non-esterified fatty acids by lipases. Hydrolytic breakdown therefore occurs during storage, but is accelerated by the more disruptive processes involved in oil extraction. Such breakdown can be arrested by brief exposure to high temperature which inactivates the lipase enzymes. The presence of significant concentrations of non-esterified fatty acids causes rancidity and spoils the quality of the final oil.

Another important function of these refining processes is to remove toxic substances that are characteristic of many plants and which are pulled out with the oil during extraction. Examples are the aflatoxins which are sometimes associated with peanut oil and which are among the most toxic substances known, and the isothyocyanates and glycosinolates of rape which poison the thyroid gland to cause goitre.

Animal carcass fats are extracted by a 'rendering' process. This involves either heating alone to dry the material and liberate the fat, or steaming in the presence of a small amount of water. The fat eventually rises to the top and can be separated. Normally, animal carcass fats do not contain the wide range of contaminating components that occur in plant oils and so need less drastic processing. As we have seen, however, adipose tissue can accumulate fat soluble contaminants such as pesticide residues and anabolic steroids used in animal production and these are likely to remain in the refined fat.

Even food fats with an image of 'naturalness', for example butter and cream, are processed by a separation procedure to enrich the fat phase compared with the original milk. The relatively simple procedures of centrifugation, churning or homogenization have little effect on the composition of milk fat, but lipases present in the milk fat globule membrane may be activated and so lead to the release of non-esterified fatty acids. Since a considerable proportion of the milk fatty acids have short chain lengths and are therefore relatively volatile, the rancid off-flavour associated with non-esterified fatty acids in dairy fats may be particularly troublesome.

Although the extraction, degumming and refining procedures described above remove those contaminants that are quantitatively most important, the resulting fats and oils, especially those from plants, may still retain small quantities of compounds that give rise to strong colours, pungent odours and unpleasant tastes. 'Bleaching out the colour' involves mixing the oil with an adsorbing substance such as Fuller's earth or activated charcoal. This normally involves contact with the adsorbent for up to 30 min at 80–90°C but there are many variations of the process, some involving more drastic temperatures. As well as removing pigments, the process may remove potentially carcinogenic polycyclic hydrocarbons if they were present in the original oil. During this process, polyunsaturated fatty acids may be oxidized to peroxides (see Fig. 3.2) and the molecules may undergo rearrangements so that the pattern of double bonds is changed from a methylene interrupted sequence (Fig. 2.2(h)) to a conjugated sequence (Fig. 2.2(j)).

'Deodorization' involves removal of the more volatile components that are responsible for unpleasant odours and tastes by high pressure steam distillation at around 240–270°C for up to 1 h. As well as eliminating flavour compounds, the process also has the advantage of removing many pesticide residues and certain mycotoxins. However, unless the steam distillation is carefully controlled, the unsaturated fatty acids are likely to suffer damage by conversion into other geometrical and positional isomers and may become polymerized to various extents. Also the processes of bleaching and deodorization reduce the content of fat soluble vitamins such as tocopherols and carotenoids, effectively reducing the nutritive value of the oil to a certain extent.

3.3.2. Heating

In some processes, such as frying, which may be used either in the industrial preparation of foods or in household cooking, fats may be subjected to elevated temperatures. During heating, some changes may occur in the structure of the fat, the nature of which depends on several factors like temperature, the length of time for which the fat is heated and the amount of air to which it is exposed. The sort of temperature involved in deep-fat frying, for example, would be about 180°C. At this temperature, the evolution of water vapour carries volatile compounds out of the fat ('steam-stripping'). The formation of a layer of volatiles and steam above the fat acts as a barrier to the entrance of too much air.

In the presence of air, the first products that are formed at lower temperatures are the hydroperoxides of unsaturated fatty acids (see

Section 3.3.7) which breakdown at higher temperatures to give a variety of oxygenated fatty acids. At higher temperatures, but without the necessity of oxygen, cyclic monomers of triacylglycerols are formed and, after that, polymeric triacylglycerols begin to accumulate (Fig. 3.3). Fats heated under household conditions may contain about 10–20% of polymerized material, but the functional properties of the oil are not noticeably worse under these conditions and such fats are not regarded as harmful (see Section 8.2.4).

3.3.3. Hydrogenation

Whether a fat is liquid or solid at ambient temperature is largely determined by the degree of unsaturation of the fatty acids in its constituent triacylglycerols. The more unsaturated they are, the lower the melting point is likely to be (Table 3.9).

Highly unsaturated oils such as are found in many seeds and in fish are unsuitable for many food fats because they have very low melting points and also because they are more susceptible to oxidative deterioration. The objective of hydrogenation, which is probably the margarine manufacturer's most important tool, is to reduce the degree of unsaturation, thereby increasing the melting point of the oil. By careful choice of catalyst and temperature, the oil can be selectively hydrogenated so as to achieve a product with precisely the desired characteristics. Indeed, the process is seldom taken to completion, since completely saturated fats, especially those that would be derived from the long chain fatty acids of fish oils, would have melting points that were too high.

Hydrogenation is carried out in an enclosed tank in the presence of

TABLE 3.9
The Melting Points of Some Isomeric Fatty Acids

Chain length	*Fatty acid*	*Melting point* (°C)
12	12:0	44·2
16	16:0	62·7
18	18:0	69·6
18	*c*-18:1	13·2
18	*t*-18:1	44·0
18	*c,c*-18:2	−5·0
18	*t,t*-18:2	28·5
18	*c,c,c*-18:3	−11·0
18	*t,t,t*-18:3	29·5

0·05–0·20% of a finely powdered catalyst (usually nickel) at temperatures up to 180°C, after which all traces of catalyst are removed by filtration. In the course of the hydrogenation, a proportion of the *cis* double bonds in natural oils are isomerized to *trans* double bonds (Figs 2.2(g) and (j)) and there is also a migration of double bonds along the chain. Hence, whereas a natural fat contains a high proportion of double bonds between the 9th and 10th carbon atoms along the chain (Fig. 2.2(f)), the products of partial hydrogenation contain double bonds that are randomly distributed. In principle, there is little difference between the chemically catalysed reduction of double bonds employed by the food industry and the enzymically catalysed reduction of double bonds by the microorganisms in ruminant animals. The main difference is that the more drastic and less controlled industrial process (the conditions used are still largely empirical) produces a wider variety of isomers and in some products a higher concentration of isomers than the biological process. *Trans* double bonds occurring in manufactured fats cannot be said to be unnatural in the sense that they do occur in nature, but they may well occur in unnaturally large concentrations. Some margarines may have up to half their unsaturated fatty acids in the *trans* configuration (Table 8.2). Tables 3.9, 8.2 and 8.3 give an idea of the changes in composition and physical properties that can be brought about by processing. It is interesting to note that the melting points of the *trans* fatty acids are considerably higher than those of their *cis* counterparts. Therefore the degree of total unsaturation can be a poor guide to the physical properties of the fat (Table 3.9).

The hydrogenation process has made it possible to extend the food uses of a number of vegetable and marine oils whose melting points would otherwise be much too low.

3.3.4. Interesterification

The nature of the fatty acids esterified in triacylglycerols is not the only factor influencing the physical properties of the fat. Another important influence is the distribution of the three fatty acids among the different positions of the glycerol molecule. We have already seen how natural fats tend to have specific asymmetrical distributions of fatty acids in the molecule. Interesterification is a method of altering the melting point of a fat by randomizing the positions of the fatty acids. Positions may be exchanged between fatty acids of the same triacylglycerol molecule (intramolecular exchange) or between fatty acids of different molecules (intermolecular exchange). The fat is heated in the presence of a catalyst (usually sodium, sodium ethoxide or sodium methoxide) to a temperature of 110–160°C. Interesterification can now be achieved more specifically with enzymes allowing new possibilities for specialized food fats. By this method, not only the physical properties but also the metabolic effects of the fat can be changed (see Section 8.2.3.).

3.3.5. Fractionation

This is a means of separating fats with different melting points. The fat is first melted and then cooled under controlled conditions. The high melting glycerides crystallize and can be filtered off.

3.3.6. Emulsification

An 'emulsion' is a stable dispersion of oil droplets in a predominantly aqueous medium or of water droplets in oil. An example of a natural oil-in-water emulsion is milk and a water-in-oil emulsion is butter. In industrial processing, to maintain the stability of an emulsion, an 'emulsifier' is required and this is normally an amphiphilic substance with polar and non-polar parts within the same molecule (see Section 2.2). Milk fat globules are stabilized by a surface layer of proteins and phospholipids. Industrial emulsification may make use of natural emulsifiers such as phospholipids (Fig. 2.1(g)) and monoacylglycerols (Fig. 2.1(f)) or synthetic emulsifiers such as polyoxyethylene monostearate. The ability to produce 'soft' margarines, in which most of the fat phase consists of unhydrogenated oils such as sunflower seed oil, depends on the development of sophisticated emulsifier technology to maintain the fat in a 'solid' form, which is still an industrial secret. The same is true for 'low calorie spreads' in which 50% of the emulsion is water.

3.3.7. Spoilage of Food Fats during Processing or Storage

The release of non-esterified fatty acids which can occur during processing or storage is often a preliminary to their oxidative breakdown. The major pathway for oxidative deterioration of food fatty acids is attack by oxygen at the double bonds of unsaturated acids with the initial formation of a peroxide (Fig. 3.2).

The more highly unsaturated the fatty acid, the greater is its susceptibility to oxidation. Peroxidation may occur by autoxidation in which the lipid catalyses its own oxidation or it may be chemically or enzymically catalysed. Haemoglobin and metals such as iron or copper are common food constituents that catalyse oxidative deterioration of fats.

A. LIPOLYTIC RELEASE OF FATTY ACIDS

$$\begin{array}{l} CH_2OCOR^1 \\ | \\ CHOCOR^2 \\ | \\ CH_2O{-}X \end{array} \xrightarrow{\text{Lipase}} \begin{array}{l} CH_2OH \\ | \\ CHOH \\ | \\ CH_2O{-}X \end{array} + R^1COOH + R^2COOH$$

B. OXIDATIVE BREAKDOWN OF FATTY ACIDS

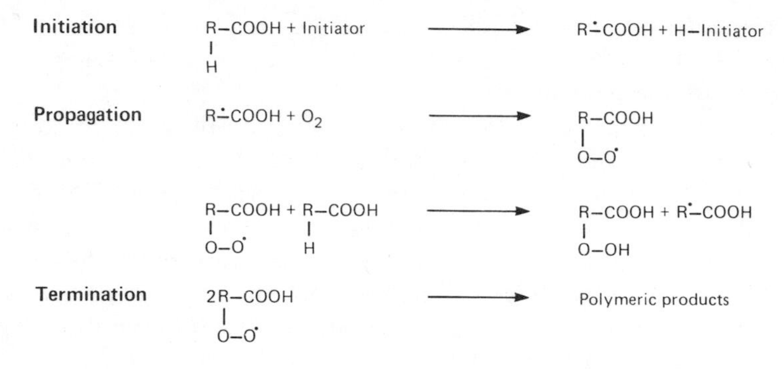

Fig. 3.2. The breakdown of lipids in tissues or foods. These processes can occur in the intact living tissue or in food when the appropriate enzymes still remain active.

(A) Illustrates hydrolytic breakdown when lipases are active, resulting in the liberation of free fatty acids. Many lipases exist, which all have somewhat different specificities for different lipid substrates. Some act only on a single type of lipid, others have broad specificity for group X.

(B) These oxidative processes may be purely chemical (autocatalytic or catalysed by metals or haem compounds) or catalysed by the enzyme lipoxygenase. The latter is a much more specific process, occurring in unsaturated fatty acids with a particular configuration of double bonds, as in linoleic acid.

The most important agent for enzymic peroxidation is the enzyme lipoxygenase, originally thought to be present only in plants, but recently shown to exist in animal tissues. The chief sources of the enzyme are in peas and beans (especially soybean), cereal grains and oil seeds. It was originally detected by its oxidation of carotene and has been used extensively in the baking industry for bleaching carotenoids in dough.

The free radicals produced during peroxide formation (ROO$^•$, RO$^•$, OH$^•$, etc., where R is a hydrocarbon chain) react at random with surrounding molecules by removing hydrogen atoms. This paves the way for a variety of addition reactions which cause damage to proteins, including enzymes, other lipids and vitamins. Vitamin A is particularly susceptible to peroxidation damage. Compounds that react quickly with free radicals — antioxidants — such as the naturally occurring tocopherols (vitamin E) or synthetic antioxidants, such as butylated hydroxytoluene (BHT), are useful in retarding peroxidation damage. As well as resulting in physical deterioration of the food and generating potentially toxic compounds, peroxidation reduces the organoleptic properties of the food. A chemical test, the peroxide value, is used to determine the level of peroxides in food (see Section 4.4.4.3).

3.4. SUMMARY

In this chapter are described the different sources of fat in the diet and how the composition of the fat varies with source. Visible fats in the diet are derived from animal storage fats such as those of adipose tissue and milk, which provide mainly saturated and monounsaturated fatty acids esterified in triacylglycerols and from seed oils, whose triacylglycerols are frequently but not exclusively rich in polyunsaturated fatty acids. Milk and coconut oil provide short and medium chain saturated fatty acids, palm oil provides longer chain saturated acids, while fish oils provide highly unsaturated very long chain fatty acids. The fatty acid composition of the storage fats of simple-stomached animals is readily amenable to manipulation by diet; that of ruminants less so.

Hidden fats are derived mainly from the muscle tissue of farm animals and from green leaves. They are characterized by a high polyunsaturated fatty content.

Cholesterol is a constituent of most animal storage fats and of structural fats but not of plant oils. Fat soluble vitamins and pigments are also present to different extents in most food fats.

Polyunsaturated fats are readily susceptible to oxidative breakdown which leads to a deterioration of food quality and demands care in processing and food preparation. The physical properties, and therefore the range of food use of fats can be modified by processing techniques such as hydrogenation, fractionation, emulsification and fractionation. These processes may also influence the nutritive value favourably or adversely.

BIBLIOGRAPHY

Brisson, G. J., *Lipids in Human Nutrition: An Appraisal of Some Dietary Concepts*, 1981, MTP, Lancaster. (Chapter 1 gives a useful account of the structures and properties of fats found in foods. Chapter 2 gives valuable information on the technology of edible fats and oils.)

Fox, P. F. (Ed.) *Developments in Dairy Chemistry — 2: Lipids*, 1983, Applied Science Publishers, London. (Contains chapters on the structure of milk lipids and manipulation of composition by nutritional means, and also on hydrolytic and oxidative rancidity associated with milk lipids. The final chapter is concerned with nutritional aspects of lipids.)

Gunstone, F. D. and Norris, F. A., *Lipids in Foods*, 1983, Pergamon Press, Oxford.

Gurr, M. I., The biosynthesis of triacylglycerols. In: *The Biochemistry of Plants*, Vol. 4, eds P. K. Stumpf and E. Conn, 1980, Academic Press, New York, pp. 205–48. (Contains information on the composition of seed oils and the structures of seed oil lipids as well as the biochemistry of seed lipids.)

Gurr, M. I., Agricultural aspects of lipids. In: *The Lipid Handbook*, 1984, Chapman and Hall, London. (Gives a concise account of the effects of agricultural practice on both animal and plant fats and some indications of future trends.)

Gurr, M. I. and James, A. T., *Lipid Biochemistry: An Introduction*, 3rd Edn, 1980, Chapman and Hall, London. (Contains a detailed account of lipid structures and properties.)

Hitchcock, C. and Nichols, B. W., *Plant Lipid Biochemistry*, 1971, Academic Press, New York. (A detailed reference work on all aspects of plant lipids: composition and structure as well as metabolism.)

McDonald, I. W. and Scott, T. W., Foods of ruminant origin with elevated content of polyunsaturated fatty acids, *World Rev. Nutr. Dietetics*, 1977, **26,** 144–207. (Gives a detailed account of the attempts, especially in Australia, to modify the composition of ruminant fats by feeding protected fats in the diet.)

Oldham, J. D. and Sutton J. D., Milk composition and the high yielding cow. In: *Feeding Strategy for the High Yielding Dairy Cow*, EAAP Publication No. 25, eds W. H. Broster and H. Swann, 1979, Granada Publishing, London, pp. 114–47. (Discusses both nutritional and non-nutritional factors that influence the yield and composition of milk fat (and other milk constituents).)

Paul, A. A. and Southgate, D. A. T., *McCance and Widdowson's The Composition of Foods*, 1980, HMSO, London. (An essential reference work for those needing data on food composition. Contains tables giving total fat, fatty acid composition, fat soluble vitamins and cholesterol.)

Wiseman, J. (Ed.), *Fats in Animal Nutrition*, 1984, Butterworths, London. (A collection of papers presented at the Annual Easter School of the University of Nottingham School of Agriculture. Contains a wealth of information on the fat composition of different farm animals and how they may be modified by agricultural practice. There are chapters on plant lipids, lipids of ruminants, pigs, poultry and fish.)

Chapter 4

Determination of the Amounts and Types of Fats in Foods

4.1. INTRODUCTION: WHAT DO WE NEED TO KNOW AND WHY?

It is not absolutely essential to an understanding of the nutritional aspects of food fats to have a detailed knowledge of lipid analysis. There are many excellent books to which the nutritionist with a more specific interest in the subject can turn and some are listed in the Bibliography to this chapter. There are two good reasons, however, why a book on fat nutrition should contain a brief description of analytical procedures. Firstly, the development of techniques since the 1950s has been enormous and without the ability to separate and identify individual lipids, which always occur naturally in complex mixtures, present knowledge of their metabolic and nutritional roles would have been impossible. Secondly, a number of issues in modern nutrition (which will be discussed further in Chapters 5, 7 and 8) are very much dependent, for their successful resolution, on the ability to quantitate, easily and accurately, the total fat content of a food or of the whole diet and to determine the amounts of different types of fats present. The issues referred to include: How much fat, ideally, should there be in the diet? How much saturated fat, *trans* fatty acids, essential fatty acids, cholesterol or fat-soluble vitamins are there in a food and how can I choose my diet to ensure an intake of enough or not too much, as the case may be?

To be able to make the appropriate food choices consumers require education in the nature and distribution of food fats, which is what this book is all about. Today, a bewildering variety of processed foods is available and even the best educated cannot be expected to have a working knowledge of their fat composition. An important aid to food selection is sensible labelling of products and to supply this the manufacturer needs at his command an appropriate analytical facility. It is important to decide precisely what information is required because although modern technology is capable of providing extremely detailed analytical data for individual lipid components, the consumer seldom requires this degree of detail. The provision of extensive analytical data would certainly add to the cost of the product, so that it is important to distinguish between the basic information that is useful to consumers and the more detailed data that may be required by the research nutritionist oı hospital dietitian.

The former requires rather basic techniques appropriate for routine use in a factory setting, not the sophisticated analyses found in the research laboratory. The practical nutritionist or dietitian may find all the compositional data required in food tables, such as found in *McCance and Widdowson's The Composition of Foods* (see Bibliography) but some research work may require direct analysis of foods as they are presented on the plate. The purpose of this chapter is to give the reader with predominantly nutritional interests a broad insight into how food fats may be analysed and the problems that are involved. In broad outline, the first task is to extract the fats from the food, making use of their greater solubility in organic solvents that do not mix with water, thereby immediately effecting a separation of the fat from all water soluble components. The extracted fat will be a complex mixture. It may be possible to determine directly some of the individual components of the mixture by chemical techniques that distinguish one lipid from another. Alternatively it may be useful or necessary to separate out the individual components before determining them chemically. Modern separation techniques can often be combined directly with a quantitative determination of each separated component thereby providing simultaneous and rapid analytical data on the types and amounts of fat components present (Fig. 4.1).

4.2. EXTRACTION AND MEASUREMENT OF TOTAL FAT IN FOODS

The first task in the analysis of food for fats is to extract the fat with a solvent. Hydrophobic fats like triacylglycerols and cholesterol are generally present in the form of large globules and are readily extractable with most solvents, including hydrocarbons such as light petroleum. Amphiphilic lipids like the phospholipids and glycolipids are closely

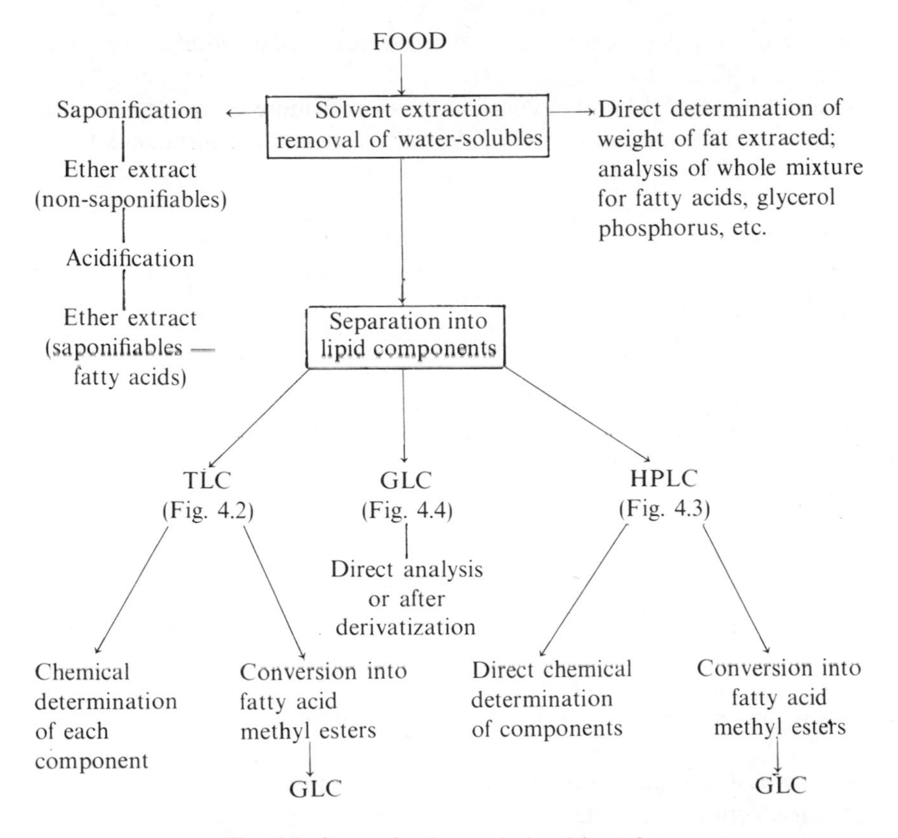

Fig. 4.1. Stages in the analysis of food fats.

associated with proteins and water, contributing to the structure of the food. Hydrocarbon solvents do not extract these lipids efficiently so that two alternative procedures need to be employed. Either the chemical bonds between the lipids and proteins must be broken first by a preliminary acid or alkaline treatment, after which a hydrocarbon will efficiently extract the liberated fat, or a more hydrophilic substance, such as an alcohol, can be introduced into the extracting solvent, obviating the need for pretreatment with acid or alkali. Fats that contain fatty acids can be separated effectively from others such as sterols and carotenoids by heating with alkali. This is the practice used in soap-making and is known as saponification. The 'non-saponifiables' (steroids, carotenoids, etc.) are extractable into diethyl ether from the alkaline solution, whereas the fatty acids are not extractable until the mixture has been acidified.

Thus, the method of extraction very much depends on the nature of the food and its constituent fats. Four main methods of fat extraction are used in food chemistry and are now described.

4.2.1 Soxhlet Method for 'Extractable' Fat

In this method the fat is extracted with light petroleum from the dried residue obtained during the determination of the moisture content of the food. The extraction is achieved by continuous refluxing of the solvent over the finely-divided dried food for a period of several hours on a water or steam bath. After extraction, the solvent is removed by evaporation and the residue is weighed. The method will only determine readily extractable hydrophobic fats and is therefore rather less precise than the other methods described here.

4.2.2. Weibull Method for Total Fat

This method is suitable for most wet proteinaceous foods, except those with a significant carbohydrate content such as dairy products. The sample is boiled in dilute hydrochloric acid to disrupt the chemical bonds binding the lipids into the structure of the food. After acid treatment the liberated fat is extracted with light petroleum.

4.2.3. Roese–Gottlieb Method for Fat in Milk Products

In this method, the sample is treated with a solution of ammonia in alcohol to disrupt lipid protein bonds without breaking down heat- and acid-sensitive carbohydrates which would interfere with the extraction. It is also useful for products in which partial glycerides (mono- and diacylglycerols) are present in addition to triacylglycerols. After pretreatment, the liberated fat is extracted with a mixture of light petroleum and diethyl ether.

4.2.4. Chloroform–Methanol Extraction

This method is applicable to almost all types of food and is the method of choice when the analyst wishes to proceed with the analysis of individual fat components in the mixture.

The fat is extracted from the sample by vigorous stirring with a chloroform–methanol mixture. There are several procedures in use but a very detailed account for food chemists can be found in Osborne & Voogt's book (see Bibliography). Because water is present in the sample and a calculated amount is added during the procedure, the extract separates into two phases after the solid residues have been filtered off. The chloroform constitutes the lower phase and contains the extracted fats. The upper phase is aqueous methanol, containing water soluble compounds. The chloroform layer is washed with water (containing salts to prevent emulsion formation) and dried over a desiccant such as anhydrous sodium sulphate. The total fat residue can be weighed after evaporation, then redissolved for further analysis as described below. Where unsaturated fats are present, it is advisable to guard against oxidation by including an antioxidant (e.g. butylated hydroxytoluene) in the extracting solvent and/or evaporating the solvent in the final stage under a blanket of inert gas. With many plant foods, there is a danger that lipases could be activated during the extraction procedure causing a breakdown of the lipids into free fatty acids (Fig. 3.2). This is best prevented by prior extraction with isopropanol which inactivates the lipases.

4.3. MEASUREMENT OF THE ENERGY VALUE OF FATS

The importance of the energy value of fats is discussed in Chapter 5 and for definitions of energy units and detailed discussion the reader should refer to the next chapter (Section 5.2.2). For most purposes it will be sufficient to estimate the energy value of a food fat by multiplying the weight of fat in the food by an average factor for the energy value of fat per gramme which is $37\,kJ\,g^{-1}$ (or $9\,kcal\,g^{-1}$). The energy value of a food or its fat component can be measured directly by bomb calorimetry. The sample is burned in an atmosphere of oxygen and the heat generated by this process is measured and used to calculate the energy value of the sample. Direct measurements of this kind may be necessary in some detailed research work but in general the application of the energy factor suffices.

4.4. DETERMINATION OF INDIVIDUAL FAT COMPONENTS

4.4.1. Direct Chemical Determination in the Mixture

The major fat components of foods can be determined chemically in the total mixture by making use of specific chemical features of each type of fat. Thus, phospholipids are characterized by their phosphorus content which can be measured by the blue colour that is formed when the complex formed between inorganic phosphate and molybdate ions is treated with a suitable reducing agent. The fat must first be digested with perchloric acid to liberate inorganic phosphate. Cholesterol can also be determined by converting into a derivative that is coloured. The intensity of the colour is related to the concentration of the compound in solution and is measured in a colourimeter or spectrophotometer. Glycolipids can be estimated by chemical reactions of the sugars released by acid hydrolysis. Triacylglycerols are less amenable to this approach since they have no characteristic feature to distinguish them from other lipids. Alkaline hydrolysis yields only fatty acids and glycerol. The latter can be determined by a colour reaction or an enzymic method but this would also determine glycerol from other glycerolipids present in the mixture. The modern approach would normally be to determine triacylglycerols in terms of their fatty acid or glycerol content after separation from other lipids in the mixture.

4.4.2. Separation Techniques

Before the advent of chromatography it was necessary to employ techniques such as distillation, crystallization and urea fractionation (see Gunstone & Norris in the Bibliography). These methods require relatively large amounts of material and are still valuable in the commercial scale preparation of fats, but for separation prior to quantitative analysis, chromatographic techniques are now universally employed.

4.4.2.1. Adsorption Chromatography on Silica

A 'chromatograph' consists of two separate chemical phases, one stationary and one moving. The chemical components in the mixture applied to the chromatograph distribute themselves between the moving and stationary phases according to their physical–chemical properties (e.g. molecular weight, hydrophobic versus hydrophilic nature, degree of unsaturation, etc.). By careful choice of stationary and moving phases, the temperature and the time of operation of the chromatograph, excellent separations can be achieved even when the differences in physical properties between components of the mixture are exceedingly small (e.g. difference in one carbon atom between chain lengths of fatty acids or a difference in geometrical configuration of one double bond). It is usually possible to provide a visual record of the separated components and this is known as a 'chromatogram'.

The stationary phase material with the widest application is silica. It can be arranged as a thin layer on a sheet of glass or plastic. Small quantities of fats dissolved in a solvent are applied to the layer near the base of the plate, which is then placed in a shallow trough of solvent so that the solvent percolates up through the layer by capillary action. Components of the mixture move up with the solvent at different rates and are thus separated from each other. Their positions can be detected by spraying the plate with a reagent which reacts with the components to give a colour. Compounds can be identified by running known standards on the same plate. They can be quantified by eluting them from the plate

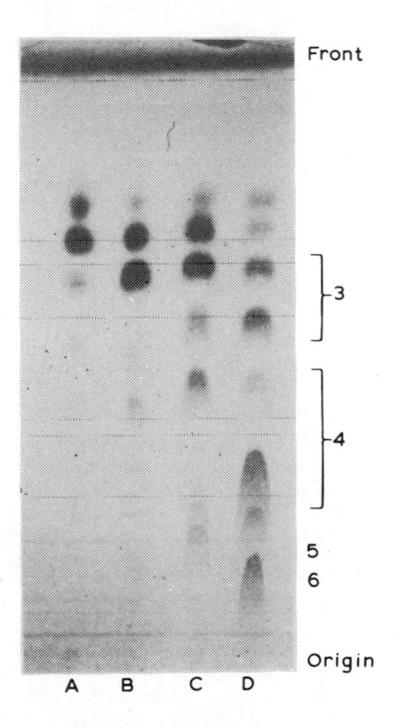

Fig. 4.2. An example of a thin layer chromatogram. This illustrates TLC of a natural mixture of triacylglycerols on silica gel G impregnated with silver nitrate (5% w/w). Solvent: isopropanol–chloroform, 1·5:98·5 (v/v). Spots were located by spraying with 50% sulphuric acid and charring in an oven. A, Palm oil; B, olive oil; C, peanut oil; D, cottonseed oil. The numbers along the side represent the total number of double bonds in the triacylglycerol molecules in each separated fraction. (Reproduced from M. I. Gurr and A. T. James, *Lipid Biochemistry: An Introduction*, 3rd Edn, 1980, Chapman and Hall, London, by kind permission of the publishers.)

with a solvent and determining them chemically with specific reactions, as indicated in Section 4.4.1. Alternatively it may be possible to measure the intensity of the colour on the plate and relate this to the quantity present. This is potentially valuable but not widely applicable because of technical difficulties of achieving a direct relationship between intensity and quantity. This technique is known generally as 'thin layer chromatography' (TLC) (Fig. 4.2).

More sophisticated separations on thin layer plates can be achieved by incorporating into the silica a substance that interacts with specific chemical groupings. Thus, the incorporation of silver nitrate (argentation TLC) allows the separation of fatty acids according to the number or configuration (*cis*/*trans*) of double bonds in fatty acid chains.

It is also possible to arrange the silica in the form of a column of material packed in a thin glass tube (column chromatography). The sample is applied to the top of the column and the solvent is run through continuously. Eluted components appear at intervals and must be detected in some way and then either collected for further analysis or passed into a device for continuous on-the-spot analysis.

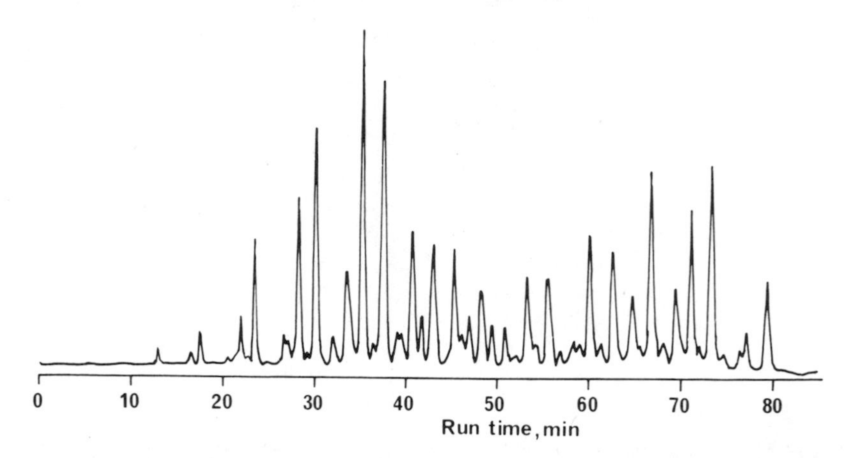

Fig. 4.3. An example of a high performance liquid chromatogram. This illustrates the separation of the triacylglycerols in butter on a 250 mm × 46 mm column of Spherosorb ODS 2 (5 μm). The solvent is acetone in acetonitrile, the composition changing in a linear manner from 15% acetone at 0 min, 20% at 5 min, 45% at 35 min to 75% at 85 min. The chromatograph is programmed to print out retention times and areas of each component peak which can be identified with standard compounds or by chemical and physical analysis. (Reproduced by kind permission of Dr J. L. Robinson, Food Science Department, University of Reading.)

```
Operator:JLR
Method:BUT3
Detector:MDII ATIN1 PMS2 ES25 AIR22PSI
Column:#44 [illegible] 9/83 WITH 5CM GUARD
Flow:1.5
Gradient file:BUT6
Mobile Phase:ACETONE IN ACETONITRILE
Inject amt.:20 MICROLITRES
Comments:85%@0M, 80%@5M, 55%@35M, 25%@85M
        :POST REPAIR CHECK ON MDII
        :

Run time: 85.00
Slope Sensitivity:  0.05
Peak width:  1.00
Min. area:      2500.
```

RT	Area	Area%	Label
5.25	6109.	0.153	--
9.48	3166.	0.079	--
10.24	412[illegible].	0.103	--
12.92	6428.	0.161	--
13.22	5058.	0.127	--
1[illegible].[illegible]4	872[illegible].	0.219	--
17.[illegible]4	29006.	0.727	--
21.85	45241.	1.133	--
23.35	102255.	2.562	--
26.61	41556.	1.041	--
28.11	176[illegible]4.	4.431	--
29.18	4165.	0.104	--
29.94	252586.	6.328	--
31.88	28517.	0.714	--
33.43	145927.	3.656	--
35.06	385489.	9.658	--
36.47	22997.	0.576	--
37.44	377187.	9.450	--
39.08	32851.	0.823	--
39.50	32684.	0.819	--
40.72	124539.	3.120	--
41.69	39915.	1.000	--
43.07	173421.	4.345	--
45.31	132245.	3.313	--
46.20	38970.	0.976	--
46.98	48692.	1.220	--
48.30	109309.	2.738	--
49.45	40908.	1.025	--
50.85	29017.	0.727	--
53.30	108345.	2.714	--
54.25	20388.	0.511	--
55.53	143161.	3.587	--
56.93	17032.	0.427	--
58.33	11486.	0.288	--
59.03	7544.	0.189	--
60.18	132069.	3.309	--
62.62	142396.	3.567	--
64.86	64387.	1.613	--
66.79	217313.	5.444	--
68.[illegible]	[illegible]	0.7[illegible]	
69.55	[illegible]	2.2[illegible]2	
71.12	143021.	3.5[illegible]3	--
71.97	11425.	0.2[illegible]6	
73.38	236661.	5.929	--
74.63	8181.	0.285	--
76.56	1771[illegible].	0.[illegible]	
77.29	34482.	0.8[illegible]4	--
77.95	3593.	0.0[illegible]	-
79.56	104146.	2.609	--

Fig. 4.3—*contd*. The printout.

The technique which is now beginning to dominate all others in fat separation and analysis is 'high performance liquid chromatography' (HPLC). It is quicker and more efficient than older methods of column chromatography and is particularly useful for hydrophobic lipids and fat soluble vitamins. The loading of the samples onto the column is automated, the programming of the rate of flow and composition of the mobile phase through the column is computer controlled and the effluent is passed through a detector linked to a printer which records the amounts and identities of the components (Fig. 4.3).

4.4.2.2 Gas-liquid Chromatography (GLC)

In this technique, the stationary phase is a liquid either adsorbed onto an inert solid packed into a long glass column (packed columns) or adsorbed onto the glass surface of the column itself (capillary columns). The moving phase is a gas. Like HPLC, this technique is highly automated and modern systems are computer-controlled with a final print-out of the complete analysis in terms of quantities and identities of the components. Its use is limited by the volatility of the fats to be separated and its main use is for the separation of fatty acids. These are removed from the complex lipids by chemical hydrolysis and transformed into their methyl esters for analysis by GLC (Fig. 4.4). More recently, however, it has been successfully applied to the separation of complex mixtures of triacylglycerols.

4.4.3. Identification of the Separated Components

As implied in previous paragraphs, separated components may be identified either by collecting them from the chromatograph and applying specific chemical or enzymic techniques or by comparing their relative positions on the chromatogram with authentic chemical standards. If further characterization is required, it can be obtained by feeding the compounds that are eluting from the chromatograph into a mass spectrometer. The mass spectrum can precisely identify even the most complicated of chemical structures.

4.4.4. Methods Appropriate for Different Classes of Food Lipids

4.4.4.1. Triacylglycerols

Triacylglycerols can be readily separated from other lipid classes by TLC. The amounts present can be measured by adding a known weight of a fatty acid which is not expected to be present in the mixture (margaric acid, 17:0, is often chosen as it is rarely found in food fats).

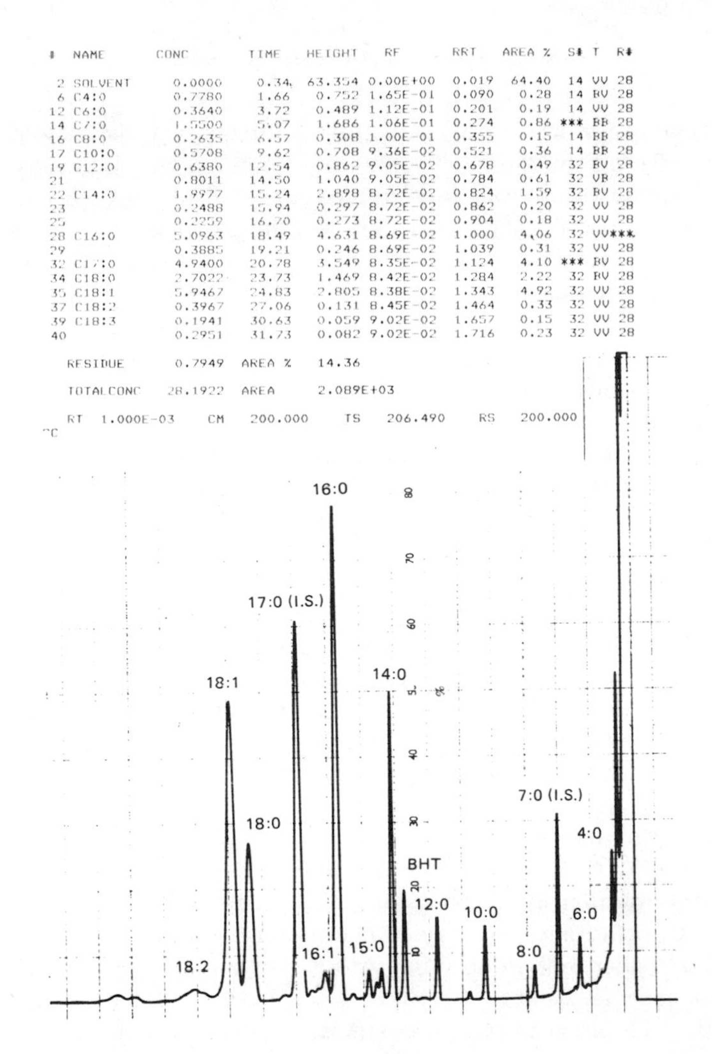

#	NAME	CONC	TIME	HEIGHT	RF	RRT	AREA %	S#	T	R#
2	SOLVENT	0.0000	0.34	63.354	0.00E+00	0.019	64.40	14	VV	28
6	C4:0	0.7780	1.66	0.752	1.65E-01	0.090	0.28	14	BV	28
12	C6:0	0.3640	3.72	0.489	1.12E-01	0.201	0.19	14	VV	28
14	C7:0	1.5500	5.07	1.686	1.06E-01	0.274	0.86	***	BB	28
16	C8:0	0.2635	6.57	0.308	1.00E-01	0.355	0.15	14	BB	28
17	C10:0	0.5708	9.62	0.708	9.36E-02	0.521	0.36	14	BB	28
19	C12:0	0.6380	12.54	0.862	9.05E-02	0.678	0.49	32	BV	28
21		0.8011	14.50	1.040	9.05E-02	0.784	0.61	32	VB	28
22	C14:0	1.9977	15.24	2.898	8.72E-02	0.824	1.59	32	BV	28
23		0.2488	15.94	0.297	8.72E-02	0.862	0.20	32	VV	28
25		0.2259	16.70	0.273	8.72E-02	0.904	0.18	32	VV	28
28	C16:0	5.0963	18.49	4.631	8.69E-02	1.000	4.06	32	VV	***
29		0.3885	19.21	0.246	8.69E-02	1.039	0.31	32	VV	28
32	C17:0	4.9400	20.78	3.549	8.35E-02	1.124	4.10	***	BV	28
34	C18:0	2.7022	23.73	1.469	8.42E-02	1.284	2.22	32	BV	28
35	C18:1	5.9467	24.83	2.805	8.38E-02	1.343	4.92	32	VV	28
37	C18:2	0.3967	27.06	0.131	8.45E-02	1.464	0.33	32	VV	28
39	C18:3	0.1941	30.63	0.059	9.02E-02	1.657	0.15	32	VV	28
40		0.2951	31.73	0.082	9.02E-02	1.716	0.23	32	VV	28

RESIDUE 0.7949 AREA % 14.36

TOTALCONC 28.1922 AREA 2.089E+03

RT 1.000E-03 CM 200.000 TS 206.490 RS 200.000

Fig. 4.4. An example of a gas-liquid chromatogram. This illustrates the separation of fatty acid methyl esters by GLC on a 2 m × 4 mm column of 10% polyethylene glycol adipate on Diatomite C-AAW. The oven temperature was programmed from 65 to 185°C at 8°C min^{-1} in order to separate short chain volatile fatty acids and long chain less volatile fatty acids in the same run. The sample consists of fatty acids derived from a milk diet for calves. Heptanoic and heptadecanoic acids were added as internal standards. By injecting known quantities of these substances (which are not expected to be present in the natural mixture) the concentrations of unknown components can be calculated. This is done by a computer which is programmed to print out a list of analysed components, their identities and concentrations in the mixture, as illustrated at the top of the figure. (Reproduced from M. I. Gurr and A. T. James, *Lipid Biochemistry: An Introduction*, 3rd Edn, 1980, Chapman and Hall, London, by kind permission of the publishers.)

This is known as an 'internal standard'. All the fatty acids, those present in the triacylglycerol plus the internal standard are then converted chemically into their methyl esters and analysed by GLC. Because the amount of internal standard was known, the amounts of all other fatty acids can be calculated. The total weight of these acids can then be used to calculate the weight of triacylglycerols from which they came using a conversion factor. This method has the advantage of measuring not only the total weight of triacylglycerols but also of identifying the fatty acids present and their relative proportions.

Triacylglycerols can also be separated and determined quantitatively by GLC but HPLC is now becoming the method of choice. This method will not, of course, give any direct information about the individual fatty acids present.

4.4.4.2. Phospholipids

Good separation can be achieved by TLC or with increasing success by HPLC. Individual components can be degraded chemically to liberate constituent parts of the molecules: the fatty acids, glycerol, phosphorus and base (see Fig. 2.1(g)) which can be analysed to provide detailed information about the nature of the phospholipids present in the mixture.

4.4.4.3. Fatty Acids

Although HPLC has been used to separate individual fatty acids, GLC is the most widely used method and will probably continue to be so. Excellent separations of complex mixtures according to carbon chain length and degree of unsaturation can be achieved on packed columns. Using capillary columns, which however are not so simple to operate in a routine laboratory as packed columns, sophisticated separations of *cis* and *trans* isomers and positional isomers can be achieved. This may be quite important in assessing the extent of modification of natural oils by industrial processing or for assessing the extent of accumulation of abnormal isomers in a diseased tissue. For most laboratories, with less sophisticated instrumentation, the surest method for determining isomeric fatty acids is *a priori* separation by argentation TLC into classes defined by the number and geometrical configuration of the double bonds (i.e. saturated, monoenoic, dienoic, etc., and *cis* and *trans* isomers). This system does not separate according to chain length but each class of fatty acids can be eluted from the TLC plate with a solvent and the mixture in the eluate can be further resolved by GLC.

When the mixture being analysed is either relatively simple or well characterized, the fatty acids can be identified by their positions on the chromatogram relative to a reference fatty acid, usually palmitic acid. If there is any doubt or the mixture is particularly complex, the identities of the fatty acids should be checked either by running on another chromatographic system with different separating characteristics or, more rigorously, by chemically identifying each component. Today, mass spectrometry would be the method of choice.

When foods are poorly stored, there is a distinct possibility of oxidative deterioration, expecially affecting the polyunsaturated fatty acids. Among the products formed in this process are peroxides of unsaturated fatty acids (see Section 3.3.7) and the total extent of peroxide formation can be determined by a chemical test. The result, the 'peroxide value', is usually expressed in milliequivalents of peroxide per kilogram of fat and it is often specified that fresh fats should have a peroxide value of 1·0 or less. For research purposes it may be necessary to determine the individual oxidized components which can be done by GLC or HPLC.

4.4.4.4. Sterols

For the nutritionist or food scientist, it is rarely necessary to separate, identify and estimate individual sterols because cholesterol is so overwhelmingly the predominant sterol in diets. Plant sterols are, of course, present in most diets but are hardly absorbed and contribute little to metabolism. Cholesterol can be determined chemically by a reaction that results in the formation of a colour, or by an enzymic technique. It is also possible to use GLC to determine cholesterol quantitatively if a known amount of an internal standard (e.g. 7-α-cholestane) is added to the food before lipid extraction).

4.4.4.5. Fat-soluble vitamins

Vitamin A: The modern method of choice for analysing vitamin A, β-carotene and other precursors with vitamin A activity is automated HPLC. For laboratories not equipped for this purpose, the older method is still very useful and requires little sophisticated equipment. The sample is saponified and the unsaponifiable lipids extracted with diethyl ether. Vitamin A can be separated from carotenoids by column chromatography on alumina and carotenoids can be further separated from each other by magnesia chromatography. Because these compounds are coloured they can be readily determined by spectrophotometry.

Vitamin D: Normally, vitamin D is determined by GLC after converting it chemically into a suitable derivative. The food is saponified and the unsaponifiables extracted with diethyl ether. Sterols such as cholesterol can be removed by precipitation with digitonin. Vitamin D is then converted into a derivative which is susceptible to analysis by GLC. This method is ideal for foods, such as marine oils, which contain more than 1 $\mu g\, g^{-1}$. It is less useful for milk products and other foods which contain small quantities of the vitamin and relatively large amounts of interfering substances. HPLC methods are now being developed and will probably be the method of choice in the future.

Vitamin E: Biologically active vitamin E (mainly α-tocopherol) is usually separated from other components by TLC. The compound is eluted and converted chemically into a derivative whose concentration can be measured spectrophotometrically. More recently, efficient HPLC separations have been developed and this will undoubtedly be the method of choice in the future.

4.5. SUMMARY

In determining the amount of fat present in a food, use is made of the solubility of fats in organic solvents rather than in water. The solvents and extraction conditions have to be tailored to the type of fat and the way it is bound into the structure of the food. Crude fat can be determined directly by weighing the residue after evaporating the extraction solvent. More detailed determination of individual fats in the mixture has to rely on a family of separation techniques known generally as chromatography. The aim of modern analytical methods is to determine the identities and quantities of components of a fat mixture by a single automated technique.

BIBLIOGRAPHY

Christie, W. W., *Lipid Analysis*, 1983, Pergamon Press, Oxford. (One of the most comprehensive treatises on the analysis of lipids in foods and in biological tissues.)

Gunstone, F. D. and Norris, F. A., *Lipids in Food*, 1983, Pergamon Press, Oxford.

Gurr, M. I. and James, A. T., *Lipid Biochemistry: An Introduction*, 3rd Edn, 1980, Chapman and Hall, London. (Chapter 1 gives a general account of lipid extraction and principles of analysis. The analysis of specific lipid classes is described briefly at the end of each chapter.)

John, A. R. and Davenport, J. B. (Eds), *Biochemistry and Methodology of Lipids*, 1971, John Wiley and Sons, New York. (Despite its age, contains some useful methods for specialized applications which are often omitted from general reference works.)

Osborne, D. R. and Voogt, P., *The Analysis of Nutrients in Foods*, 1978, Academic Press, London. (A comprehensive account of food analysis. The first part discusses the principles of food analysis; the second part is a very practical step by step account of analytical methods.)

Paul, A. A. and Southgate, D. A. T., *McCance and Widdowson's The Composition of Foods*, 1980, HMSO, London.

Perkins, E. G. and Visek, W. J. (Eds), *Dietary Fats and Health*, 1983, American Oil Chemists Society, Champaign, Illinois. (Chapters 6–12 give interesting accounts of the applications of modern methodology.)

PART II

THE METABOLIC AND NUTRITIONAL ROLE OF FATS

Chapter 5

The Place of Fat in the Diet

5.1. THE CONSUMPTION OF FAT IN DIETS

Each one of us has an individual eating pattern which may be influenced by personal likes or dislikes, or more practical considerations of cost and availability. The range of intakes of individual nutrients or food components within a population is likely to be enormous. Some will be eating a diet in which fat makes a relatively minor contribution to energy intake compared with carbohydrates, others will obtain most of their dietary energy from fat.

On average, fat contributes about 42% of the energy in the diet of people in the UK. It is more difficult to assess the absolute amounts of fat eaten. The most reliable figures for food consumption are to be found in the *Annual Report of the National Food Survey Committee* (see Bibliography). In 1981, the average Briton consumed food with an energy value of 9·3 MJ (2210 kcal) daily at home. This contained about 103 g fat. Many people forget that these figures do not include food eaten outside the home, so that the true consumption is likely to be somewhat higher. Using trade figures which reflect the total amounts of fat available, it can be calculated that the maximum consumption could be as much as 12·6 MJ energy and 139 g fat. Allowing for some wastage of fat, the true fat intake is likely to be in the region of 120 g per person per day. Just under half is contributed by saturated fatty acids, 40% monounsaturates and 12% polyunsaturated fatty acids. It can be calculated that roughly 75% of the fat intake comes from triacylglycerols, 16% from phospholipids, 1% from cholesterol and 5% from glycolipids.

Foods have widely different fat contents and contribute different amounts of fat to the diet. Milk and dairy products account for just under 1/3 of fat intake and meat just over 1/4 (Table 5.1). Roughly three times as much fat comes from animals as from vegetable sources.

TABLE 5.1.
Major Sources of Fat in the UK Diet (per person per day)

	Milk	Butter	Cheese and other dairy products	Margarine	Other fats and oils	Carcass meat	Offal	Meat products	Poultry	Fish	Eggs	Cereal[a] products	Nuts	Total
Total fat g	12·9	12·7	9·9	17·3	9·3	7·7	0·3	11·1	1·4	1·8	2·6	9·0	1·1	97·1
Total fat % total	13·3	13·1	10·2	17·8	9·6	7·9	0·3	11·4	1·4	1·9	2·7	1·1	1·1	100
Fatty acids (g)														
Saturated	8·3	14·8 (Butter + Cheese)		9·3 (Margarine + Other fats)		3·5	0·1	4·3	0·4	0·6	0·9	4·2	0·4	48·8
cis-Mono-unsaturated	2·9	4·7		8·7		2·9	0·1	4·8	0·6	0·6	1·1	2·4	0·4	29·5
trans-Mono-unsaturated	0·6	1·0		1·7		0·3		0·3				0·8		4·7
Linoleic (18:2, *n*–6)	0·3	0·4		4·3		0·3		0·9	0·3	0·1	0·3	1·0	0·3	8·2
Linolenic (18:3, *n*–3)	0·1	0·2		0·7		0·1		0·1				0·1		1·3
Cholesterol (mg)	38	74		44		43	17	46	27	16	126	12		443

Adapted from R. Burt and D. H. Buss, Dietary fatty acids in the UK, *J. Clinical Practice*, 1984, in press.
[a] Includes baked foods made with flour, etc.

Patterns of fat consumption within the UK population have been changing steadily throughout the 20th century. It is difficult to assess precisely the food consumption at the beginning of the century, but the average individual's energy consumption was probably something like 11·5 MJ (2760 kcal) of which 32% was fat. By 1955, household consumption was 11·0 MJ (2640 kcal) with 36% of the energy as fat and by 1981, although the household consumption figure for energy had dropped to 9·3 MJ (2210 kcal) per person per day, the proportion of this which was fat was 42% (Fig. 5.1).

Ministry of Agriculture, Fisheries and Food (MAFF) figures for the last few years have indicated a trend towards falling total energy consumption but a steady proportion of energy eaten as fat. The small changes in total fat intake, however, have masked quite significant changes in the types of fat eaten. Thus, there has been a steady decline in butter consumption since the late 1970s, balanced by increasing sales of

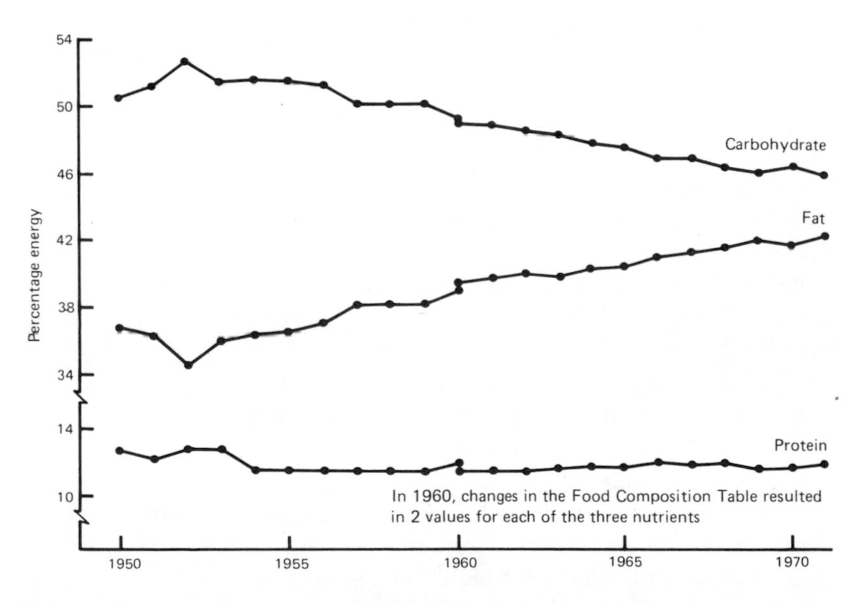

Fig. 5.1. The percentage of energy derived from the protein, fat and carbohydrate in the diet of the average household in Great Britain, 1950–71. This is the period showing the greatest change. Since 1971 there has been little change in the proportions of dietary energy derived from these three nutrients, the figures for 1981 being carbohydrate 44·9%, fat 42·2% and protein 12·9%. (Figures from MAFF, Household food consumption and expenditure, *Annual Report of the National Food Survey Committee*, 1981, HMSO, London.)

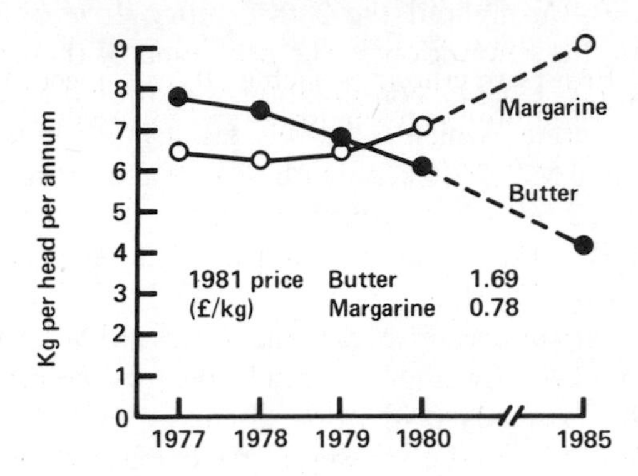

Fig. 5.2. Trends in butter and margarine consumption in the UK. The dotted line represents projections that were being made in 1981 on the basis of trends at that time. However, during 1983 there were distinct signs that the downward trend in butter consumption and the upward trend in margarine consumption were leveling out. (Figures from *Dairy Facts and Figures*, 1981, Milk Marketing Board, Thames Ditton.)

margarine (Fig. 5.2). This has in part been influenced by price and in part by the increasing interest in diet and health and the persuasion that increasing the proportion of polyunsaturated fat in the diet may be beneficial (see Chapter 8). There has indeed been a trend towards increased consumption of polyunsaturated vegetable oils and products based on them with the result that the ratio of polyunsaturated to saturated fatty acids (*P/S* ratio) in diets has increased slightly in recent years (0·17 in 1959, 0·22 in 1972 to 0·25 in 1981).

In the UK there are also interesting regional differences in fat consumption. Thus, butter consumption in Wales, south-west England and the Midlands is much higher than the national average. Margarine consumption, however, although again higher than average in the Midlands and south-west is lower than average in Wales. Fat consumption also seems to depend on family income and on the size of the family. There is a tendency for higher income groups to eat more fat (but not necessarily more total energy) than lower income groups and for less fat to be eaten per person as the number of children in a family increases, irrespective of income. There do not seem to be any significant differences between the *P/S* ratios of fats eaten by these different groups.

There are, of course, considerable national variations in patterns of food intake and the amounts and types of fats eaten. On the whole, in affluent countries like the USA, Canada, The Netherlands, West Germany, Sweden, Australia and New Zealand, the proportion of fat in the average diet is similar to that in Britain, and so is the total amount eaten. In some European countries such as Italy, Greece, Yugoslavia and Portugal, fat consumption is much lower and in Japan, the average daily fat consumption is only a little over a third of that in Britain.

5.2. WHY DO WE EAT FAT?

5.2.1. The Palatability Factor

The primary object of eating food is to stay alive and in many parts of the world today, the sole function of food is to fulfil this purpose. However, in societies fortunate enough to have an overabundance of food, eating is numbered amongst the foremost pleasures of life. Food is eaten for the pleasure it gives and eating has become a social activity rather than a necessity.

Food can be considered nutritious only in so far as it is good enough to eat and the presence of fat contributes substantially to palatability. A fat-free diet would not only be deficient in nutrients — the essential fatty acids and fat soluble vitamins — but would also be extremely unpleasant to eat. Fats contribute to palatability, principally, in two ways; by olfactory responses in the nose or mouth to lipids or lipid breakdown products or in responses to their texture in the mouth. This property of 'mouth-feel' is very complex and may exhibit the attributes of chewiness, grittiness, stickiness, smoothness or oiliness depending on the nature of the fat and its structural relationship with other components of the food.

Odour normally results from a perception of small fatty molecules that are volatile. Water solubility is an important factor in the perception of flavour and aroma because it is substances in aqueous solution that interact with taste and aroma receptors. Thus short chain fatty acids such as acetic, propionic and butyric acids have a more intense taste and smell than longer chain fatty acids because of their greater water solubility and volatility. Fats may also influence the taste of other food components by retarding their passage into saliva. Hence, caffeine is more easily tasted in water than in oil. Flavour and aroma compounds such as short chain fatty acids may influence the flavour of food in different ways according to the nature of the food. For example, the rancidity of milk is due to those same short chain fatty acids that in cheese provide the characteristic and much prized flavour.

Butyric acid is a key flavour component of butter but there are an enormous number of straight and branched chain acids that in combination contribute to the flavour and palatability of dairy products. They may arise by chemical or enzymic hydrolysis of the triacylglycerols and, indeed, lipolytic breakdown of milk fat by microorganisms is a deliberate measure in margarine making to enhance flavour. The flavour of deep fried foods is also, in part, a result of fatty acids formed by hydrolysis and of their breakdown products formed by thermal degradation.

Chemical and enzymic transformations of fats may also give rise to alcohols, esters, carbonyl compounds and lactones each of which may contribute to the unique flavours of different dairy products. As in the case of free fatty acids, the same compound may give rise to pleasant or unpleasant flavours depending on factors such as their concentration, the ratio of oil to water in the food and the proportion of other food constituents, the proteins and carbohydrates. It would be a mistake to think that only fats are responsible for flavour. The flavour of some fatty foods such as chocolate and caramelized milk may be derived from non-lipid substances formed from reactions between sugars and amino acids. Fats also contribute towards texture or mouth-feel in different ways. To swallow a pure oil is extremely unpalatable, yet emulsions like milk, butter and cream are pleasant in different ways. The mouth feel of milk is related to its colloidal structure, cream owes its attractiveness to the globular structure of the fat particles and butter has a pleasant cooling effect on the tongue. The texture and palatability of butter can be changed by adding an emulsifier such as a monoacylglycerol prior to churning or by changing its fatty acid composition by feeding ruminants 'protected' fat (see Section 3.2.1). The types of fatty acids and their distribution in the triacylglycerol molecule also affect the crystal structure of a fat and its melting range. This is especially important in relation to chocolate fats which owe their palatability to rapid melting in the mouth, yet at normal storage temperatures are quite hard and easily handled.

A crucial factor governing the contribution of fats to food structure, texture and flavour is their partition between the oil and water phases. Partition is influenced by the presence in foods of amphiphilic lipids (i.e. those containing hydrophobic and hydrophilic groups within the same molecule, for example, monoacylglycerols, phospholipids) and also by interactions with carbohydrates and proteins. Amphiphilic lipids naturally present in the food may contribute to texture (e.g. the phospholipids of the milk fat globule membrane) or the manufacturers may add natural or synthetic lipids to achieve the desired result. The development of emulsion technology has been a major contributor to the wide variety of palatable foods currently available.

5.2.2. A Question of Energy

The proteins, carbohydrates and fats in food are each capable of supplying energy: that is, the chemical energy locked up in the molecules can be transformed in the body into the energy required for movement and to provide power for the complex body chemistry. Ultimately, this energy is either deposited in the body as new protein, fat, or carbohydrate or released as heat. Indeed, Max Kleiber, in one of the most readable of scientific classics has vividly described the energy locked up in food as 'The Fire of Life' (see Bibliography).

Fats supply over twice as much energy per unit weight (38 $kJ\,g^{-1}$, 9 $kcal\,g^{-1}$) as proteins (17 $kJ\,g^{-1}$, 4 $kcal\,g^{-1}$) or carbohydrates (16 $kJ\,g^{-1}$, 3·8 $kcal\,g^{-1}$). In storing fat as the important energy reserve, animals have chosen the most efficient way. Triacylglycerols are anhydrous and represent more energy for less bulk than complex polysaccharides such as glycogen which are heavily hydrated. It is important to realize that the energy values quoted above for fats, proteins and carbohydrates are their 'gross' energy values: that is, the total amount of energy that can be released by complete combustion of the compound. This can be measured by burning the fat or other food component in oxygen in a 'bomb calorimeter' and recording the amount of heat released. In the body, this amount of energy may not be entirely available. Nutritionists therefore tend to parcel up the energy values of foods according to the extent to

---

Energy units: calories and joules. The unit of energy now used by scientists is the joule. As this is a rather small unit, nutritionists measure food energy values in thousands of joules (kilojoules, kJ) or in millions of joules (megajoules, MJ). The old unit which was superseded by the kilojoule was the kilocalorie (kcal). There are 4·18 joules for every calorie. The calorie system was of course used for a long time and people have become accustomed to assessing the energy value of food in terms of (kilo) calories. Therefore both values will be given in this book with the joule value first. It would be better, however, if everyone were to get accustomed to using the joule, since it is, scientifically, a much more logical unit. Throughout the book I shall always refer to the 'energy value' of foods. The lay term 'calorie value' is a poor term since it refers to one particular unit of measurement, now outdated.

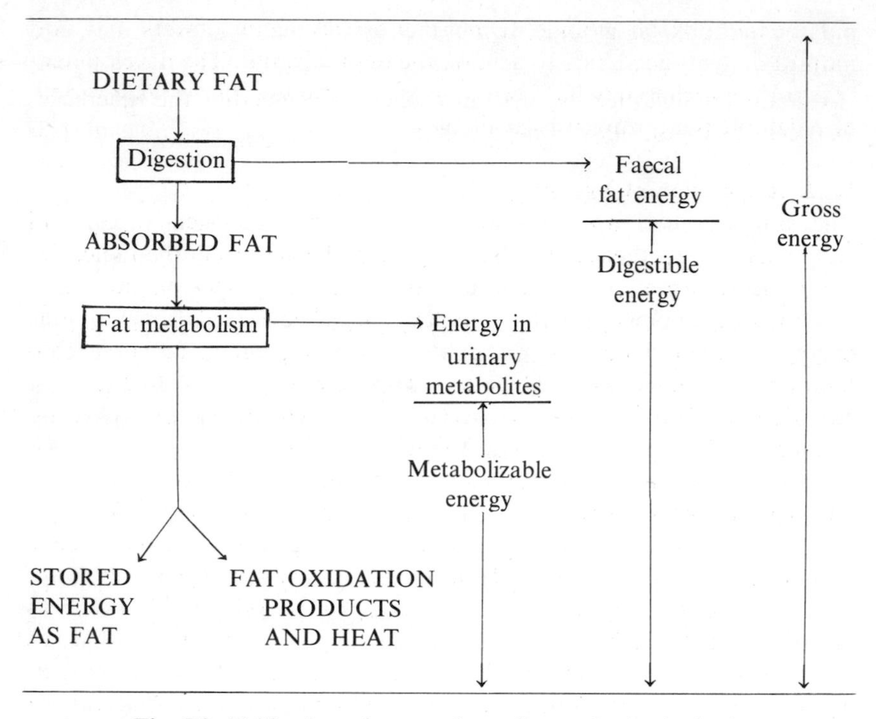

Fig. 5.3. Utilization of energy from dietary fat in the body.

which they can be used in the body (Fig. 5.3). Thus, various types of fats may be digested to different extents in the alimentary tract (Chapter 6). Extensively hydrogenated marine oils, for example, which contain a high proportion of very long chain saturated fatty acids, have a relatively low digestibility and a significant proportion of the fat (modified to some extent by bacterial action in the lower gut) will be excreted in faeces. Only the fraction that is absorbed is potentially useful to the animal and is called the 'digestible' energy. The absorbed fats then undergo extensive metabolic changes in the body in the course of which waste products may be generated which are then excreted in the urine. The energy in these waste products is lost to the body and that which remains is known as the 'metabolizable' energy.

There are small differences in the gross energy values of different fats. As the chain increases so does the gross energy per gram. Saturated fatty acids have slightly higher gross energy values than the corresponding unsaturated fatty acids. As pointed out earlier, the digestible energy tends to decrease with increasing chain length and to increase with increasing unsaturation. However, it is important to note that in healthy individuals, fat digestion is such an efficient process that the differences between the digestibility of most common food fats are insignificant. It is only when the chain length, saturation, and concentration in the diet are very high that problems may be encountered. Cholesterol is relatively poorly absorbed at high levels of inclusion in the diet, and its presence at high concentration could significantly decrease digestible energy, as could the presence of plant sterols that are not absorbed at all.

It is now possible to produce synthetic substances, not related to natural fats but with textural properties similar to food fats. These substances, like the glycerol or sucrose polyesters, are not broken down by the digestive enzymes in the alimentary tract and are therefore absorbed poorly, if at all. They therefore have a very low digestible energy and could theoretically substitute for the fats in a low energy diet. The implications are discussed in Section 8.3.3. Once absorbed, the normal food fats are metabolized very efficiently in the body, although as will be made clear in Chapters 6, 7 and 8, fats of different structures may have very different pathways of metabolism. A very good example is the contrast between the short and medium chain fatty acids found in dairy products or coconut oil and the normal long chain fatty acids. The former are absorbed directly into the blood vessels leading to the liver, where they are completely oxidized, releasing energy as heat; they are not deposited in storage fat. The latter circulate more widely around the body and may be oxidized or stored depending on the current demands for energy.

For practical purposes, the energy value of fats is of interest in relation to diet and weight control. An implication of the foregoing discussion is that the gross energy value of a fat (as would be obtained from a calorie-counting chart) may be a poor guide to its influence on body energy changes. We have to ask questions like how much fat is being eaten in the diet? How much of this is absorbed? How much of the absorbed fat is stored or wasted as heat?

In trying to understand the importance of fat (or any other food) as an energy source, it is useful to have some insight into why energy is required. Nutritionists like to think of energy requirements in terms of different compartments (Fig. 5.4). First, there is the energy for what is called basal metabolism, or basal metabolic rate. This is the energy required to maintain the essential bodily functions such as heart rate, respiration and to maintain body temperature and all the body chem-

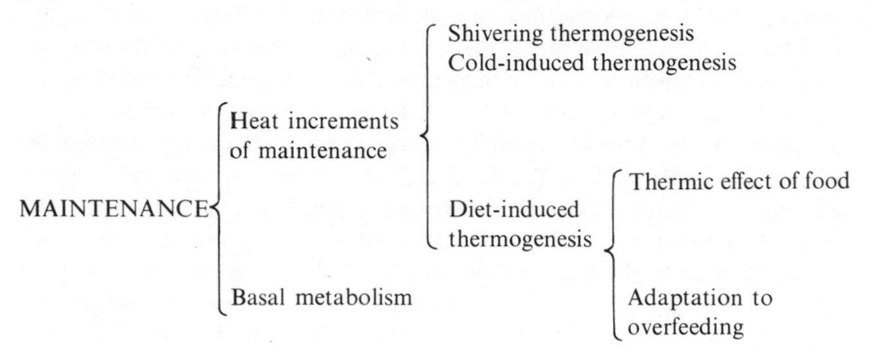

Fig. 5.4. Compartmentation of energy expenditure.

istry even when the person is completely at rest. This accounts for as much as 70% of all energy expenditure. Then energy is needed for growth and will clearly be more important for babies and children than for adults, although the energy demands for pregnancy and lactation are considerable. Physical exercise also demands energy, although, to the dismay of many slimmers, these demands are relatively small compared with basal metabolism. The body can in some respects be considered as a machine, consuming fuel (food) and doing physical and chemical work. As in any other machine, bodily processes are always less than totally efficient and the result of this inefficiency is the loss of a certain amount of energy as heat. The heat generated in response to the metabolism of food used to be known as the 'specific dynamic action' of food because it was thought to be specific for dietary protein. It is now known to be also associated with the metabolism of fat and carbohydrate as well as protein and the term 'thermic effect' of food is preferable.

Bodily processes, just like those of man-made machines, have to obey the laws of thermodynamics, so that the energy consumed in food has to be balanced by the energy used by the body; it must all be accounted for! This is the concept of 'energy balance' and an important consequence is that if the energy consumed in food is greater than is needed for the various requirements described above and in Fig. 5.4 (taking into account the heat lost during metabolism) then it must be deposited in the body stores. This is usually in the form of fat, since the ability to store glycogen is severely limited. The relationship between the energy stored and the energy consumed, however, is not a simple one because of the considerable differences there may be between the metabolic efficiency of different pathways between different individuals.

What, then, controls how much energy is consumed and how much is expended and what role does fat play in all this? Hunger, of course, plays a part in controlling food intake, but in societies where food is plentiful, it cannot be predominant. Palatability is important and as discussed earlier, fat contributes substantially to this property. Do people automatically eat more energy when they consume diets in which the proportion of fat is high compared with relatively low fat diets? The answer to this question is not known for certainty with human beings: it is difficult to do experiments that truly simulate what happens in everyday life. Rats under experimental conditions 'sense' that diets containing fat have a higher concentration of energy than their normal low fat laboratory diet and eat less food, thus maintaining a constant energy intake. They are only successful in doing this, however, when the diet contains a moderate amount of fat. Once the proportion of fat energy is more than about 30%, their ability to adjust starts to breakdown and they begin to consume more energy. The amounts that rats will eat spontaneously can be extremely large if they are 'tempted' by being given a range of foods that human beings like to eat, such as potato crisps, chocolate biscuits and the like. Experimentalists have called this the 'cafeteria' or 'supermarket' feeding system. When we use low energy substitutes as a possible aid to slimming or weight control (for example, low fat spreads which are emulsions containing about half their weight as water), we are assuming that we will be eating the same bulk and therefore less fat. This is not necessarily so and it may be that some people simply spread more on their bread and do not therefore reduce their fat consumption substantially.

The next question is: if we consume more energy by virtue of eating a lot of high fat foods (foods with a high energy density) will this extra energy necessarily be deposited as fat or does the efficiency with which fat is metabolized alter in response to increased consumption? If 'metabolic efficiency' were to decrease in response to increased food intake, the result would be an increase in heat production, which, if it had the same energy value as the extra energy consumed, would result in no increase in stored fat and no change in body weight.

Some animals do indeed adapt under certain conditions to overfeeding by decreasing their metabolic efficiency and producing more heat. This adaptive response to overfeeding is known as 'diet-induced thermogenesis' and is believed by some to exert an important control on body weight. One hypothesis suggests that the heat generated in this way originates from metabolism in a specialized tissue called 'brown adipose tissue' which normally exists to maintain body heat during cold exposure, especially in new-born animals and in hibernators. At the time of writing, the evidence that this tissue plays such a role in man, or indeed that diet-induced thermogenesis plays a key role in weight control in human beings is inconclusive. That some people are able to consume huge amounts of energy over considerable periods without apparently adding to their body stores is well documented. Especially interesting was an experiment conducted in West Germany in which subjects were asked to eat a diet with an energy equivalent of 29MJ (7000 kcal) day^{-1}, of which 85% consisted of fat. This corresponded to about 650 g fat day^{-1} (the average Briton eats about 100–120 g per day!). After periods of up to 45 days on these diets, two out of four subjects gained no weight at all. It is notable that the two people who did not gain weight were eating their fat supplements in the form of corn oil, while those who did gain some weight were eating olive oil. The authors of the paper suggested that this could mean that polyunsaturated fats were less efficiently used than monounsaturated fats, but we should treat this conclusion cautiously. The numbers of experimental subjects were very small and the result could have been due to chance. The experiment does not seem to have been repeated with human subjects; nor have similar observations been made with experimental animals, despite the large number of experiments that have been performed.

5.2.3. The Supply of Materials for Body Structure and Metabolism

Chapter 2 introduced the concepts of storage and structural fats. Dietary fats can supply structural fats in the form of fatty acids or cholesterol. A dietary supply of cholesterol and saturated or monounsaturated fatty acids is not, however, obligatory since the body can synthesize its own to compensate for dietary insufficiency as will be described in more detail in Chapter 6. Exceptions are the class of fatty acids called the 'essential' fatty acids and the fat-soluble vitamins. As none of these fatty substances can be synthesized by the body, they must be supplied in the diet, for they are essential for health in ways that will be discussed in Chapter 7.

5.3. THE SPECIAL PLACE OF MILK IN DIETS AND THE ROLES OF MILK FAT

In general, milk fulfils two functions. Firstly, it provides most of the nutrients required by the new-born from the beginning of post-natal life until weaning on to solid food. Indeed, this is the only time in life when the whole of man's nutrition is derived from one food. Secondly, some of the constituents of milk protect the new-born against its new environment. The composition of milk clearly reflects these two roles and the fat component of milk may play its part as nutrient and anti-infective agent.

The natural food for human babies is mother's breast milk, but many factors may prevent a mother from breast-feeding, in which case the substitute is a formula based generally on cow's milk. The chemical compositions of human and cow's milk are rather different as indicated in Tables 5.2 and 3.3 and the differences extend to the fat component as well as the protein.

Fat is the most variable constituent of milk, both in absolute quantity and in composition. The fat content of cow's milk depends on breed, feeding regime and on the stage of lactation. The fat content of human milk, too, depends on the stage of lactation; it tends to rise slightly from colostrum to mature milk after two weeks or so of lactation. There is also a diurnal variation in that the concentration of fat is least in the

TABLE 5.2
The Fatty Acid Composition of Milks Used in Infant Feeding ($g\,g^{-1}$ Total Fatty Acids)

Fatty acid	*Formulas based on cow's milk*	*Formulas with fat replaced by vegetable oil*	*Human milk*
4:0–8:0	3	0	0
10:0	2	0	1
12:0	3	0	5
14:0	12	trace	7
16:0	30	11	26
18:0	14	2	7
16:1	2	trace	5
18:1	31	27	37
18:2	2	58	9
18:3	1	2	3
20:4	0	0	trace

early hours of the morning. It also increases in concentration during the course of a single feed. The method of collection also influences fat content, so that milk expressed by a breast pump is richer in fat than milk dripping from the breast not being used for suckling.

A poor nutritional status of mothers results in slightly lower milk fat content, but it is not known whether the milk fat content of overweight women is higher than average. When laboratory rats are fed the 'cafeteria' diet so that they become excessively overweight, their milk contains almost twice as much fat as that of lean animals. The milk fat of obese animals contains a lower proportion of medium chain and a higher proportion of long chain fatty acids than that of lean animals (Table 5.3).

Milk fat is composed mainly of triacylglycerols present in milk as globules stabilized by a membrane of protein, phospholipids and sterols (Table 3.2). The fat globules in cow's milk tend to be larger than those in goat's or in human milk.

The fatty acid composition of cow's milk comprises more saturated fatty acids and is less affected by diet than the milk of simple-stomached animals such as human beings, because of the extensive hydrogenation of dietary unsaturated fatty acids in the rumen (Section 3.2.1). An exception is the milk of cows fed 'protected' fats which escape rumen hydrogenation and are absorbed unmodified in the small intestine. Cow's milk fat also differs from that of monogastric animals in containing very short chain fatty acids (butyric, 4:0 and caproic, 6:0) which are absent from human milk and a higher proportion of medium chain fatty acids

TABLE 5.3
Fatty Acid Composition of Milk Fat From Lean and Obese Rats (Moles Fatty Acid ml^{-1} Milk)

	Fatty acid							
	8:0	*10:0*	*12:0*	*14:0*	*16:0*	*18:0*	*18:1*	*18:2*
Lean animals	24·5	60·7	43·4	37·3	83·2	8·5	57·1	34·0
Obese animals	24·9	36·4**	28·5*	23·3*	119·9	22·5***	176·8***	65·4**

Unpublished results, by kind permission of Dr. B. A. Rolls, NIRD, Shinfield, Reading.

Asterisks indicate that the amount of a fatty acid in milk of obese animals is significantly different from that of lean animals with the following probabilities: *$p<0·05$, **$p<0·01$, ***$p<0·001$.

(caprylic, 8:0 and capric, 10:0) (Table 3.3). These fatty acids are produced in the mammary gland and are specific to milk, being produced in no other mammalian tissue. When the amount of fat in the human diet is very low, there is a marked elevation of the milk medium chain fatty acids compared with the concentration in the milk of women eating high fat diets. The proportion is highest about 8 h after the main meal and falls to lower concentrations within a few hours. Palmitic acid, the most abundant saturated fatty acid, may originate from endogenous synthesis in the gland or from the diet, depending on the relative concentrations of circulating blood sugar and lipoproteins, although what little evidence there is, suggests that in human beings diet is the most important source. Linoleic acid, the principal essential fatty acid, must originate from the diet or by mobilization from the fat stores because no animal tissue can synthesize it (see Chapter 7). The proportion of polyunsaturated fatty acids in human milk can be markedly influenced by diet. One much-quoted paper demonstrated that very high concentration of linoleic acid could occur in human milk in response to a very extreme dietary intake, but in different population groups, proportions as low as 1% in an African community living on a very low fat diet to as high as 15% in Jordanian women can be found. Despite this, the figure of 7·2% quoted in Table 3.3 is remarkably representative of samples analysed in many parts of the world. In addition to linoleic acid, human milk also contains small amounts of γ-linolenic and arachidonic acids. Whereas many authors have stressed the variability of human milk fatty acid composition (often with the implication that it might not necessarily be appropriate for the baby) in fact, human milk fat is remarkably uniform. The fatty acid pattern does not, like total fat content, vary with lactation, time of day or during the feed and the changes brought about by dietary modification are predictable. The changes in fat content are very consistent and are probably physiologically regulated to ensure that the infant receives the appropriate energy intake at any particular time.

The distribution of fatty acids within the triacylglycerol molecule in human milk tends to be rather specific, with palmitic acid occupying predominantly position 2 and the shorter and medium chain fatty acids in position 3 (Fig. 5.5). These structural considerations are probably important in relation to the absorption of milk fats by the baby.

The human baby is born before its full capacity to digest and absorb fat has developed, but by the age of one week, it can absorb more than 90% of the fat in human milk. However, at this age it can absorb only

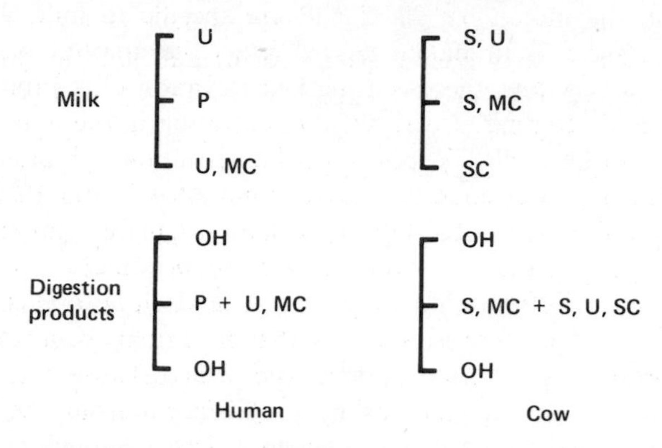

Fig. 5.5. The structures of milk triacylglycerols. The glycerol backbone of glycerides (see Figs 2.1(c) and 2.1(d)), is represented by different classes of fatty acids, represented thus; U, unsaturated fatty acids; S, saturated fatty acids; MC, medium chain fatty acids; SC, short chain fatty acids; OH, free hydroxyl of glycerol after lipolytic release of fatty acids.
The upper half of the diagram represents the intact milk triacylglycerol structures, the lower half, the products of digestion by pancreatic lipase, namely the 2-monoacylglycerols (see also Fig. 2.1(f)) and free fatty acids derived from positions 1 and 3 of the triacylglycerols. Thus from human milk the free fatty acids that are liberated are mainly unsaturated or medium chain, the long chain saturated acids remaining as 2-monoacylglycerol, while from cow's milk, the liberated fatty acids are predominantly saturated.

about 60–70% of the fat in cow's milk and less than 90% even at several months of age. The proportion of fat absorbed decreases with increasing amounts of fat in the diet, the extra fat being excreted in the faeces. In the new-born, intestinal fat digestion is limited by the amounts of digestive enzymes produced by the pancreas and the bile salts produced by the liver, since both these tissues are immature. Some fat digestion occurs in the stomach by a process that is not dependent on bile salts (see Chapter 6) and which employs digestive enzymes one of which is secreted into saliva and another that is present in breast milk.

One of several reasons for the poorer absorption of cow's milk fat is the poor solubility of free long chain saturated fatty acids. The greater proportion of stearic acid in positions 1 and 3 of the triacylglycerol molecule gives rise to higher concentrations of free stearic acid after enzymic digestion than is the case with human milk fat. Palmitic acid is predominantly esterified at position 2 of the triacylglycerol molecule in human milk, but randomly distributed in cow's milk fat (Fig. 5.5). In consequence, palmitic acid is principally found in the form of 2-monopalmitoyl-glycerol after digestion of human milk fat and is more readily solubilized in this form than as the free fatty acid. (For a detailed description of fat digestion see Section 6.2.)

Yet another factor influencing absorption of fat from cow's and human milks is the amount of calcium in the digestion products. The longer chain saturated fatty acids tend to form insoluble and poorly absorbed calcium soaps (think of the scum that forms when household soap is used in hard water) so that the combination of higher calcium and higher stearic acid concentrations in cow's milk tends to decrease the absorption of the fat.

In the production of modern baby formulas, there has been a trend towards modifying the chemical composition to be less like cow's milk and more like human milk. Thus, in respect of the fat component, some manufacturers have replaced cow's milk fat with other fats, mainly vegetable oils, with the objective of increasing the linoleic acid content. In some products, this trend has resulted in extremely high concentrations of linoleic acid, especially when sunflower seed oil has been used as the fat substitute (Table 5.2). An interesting consequence has been that in countries such as The Netherlands, when these formulas have dominated the market, the storage fat of babies has become remarkably enriched in linoleic acid (Fig. 5.6). It is not clear whether this change in body composition is of any physiological consequence, either advantageous, disadvantageous or indifferent, but some nutritionists have urged caution in marketing products whose influence on body composition can be so profound. Some milk lipids can inhibit the growth of bacteria in test-tube experiments and the virtual absence of living organisms in the stomach and small intestine of sucking rabbits has been attributed to the action of decanoic and octanoic acids released from the milk during digestion. Anti-bacterial and anti-viral activity seems to be greatest in the case of the medium chain saturated fatty acids, longer chain unsaturated acids or monoacylglycerols containing a medium chain acid. This property of fats probably derives from their ability to enter the bacterial membranes and disrupt their function. It is not known for certain whether human milk fats have special antimicrobial activity in the baby's gut, although some have made this claim. It does seem sensible, however, for baby food manufacturers to replace the fat in their formulas, not in a haphazard way, but to take into account current knowledge of the role of milk fatty acids and their special distribution in the triacylglycerol molecule.

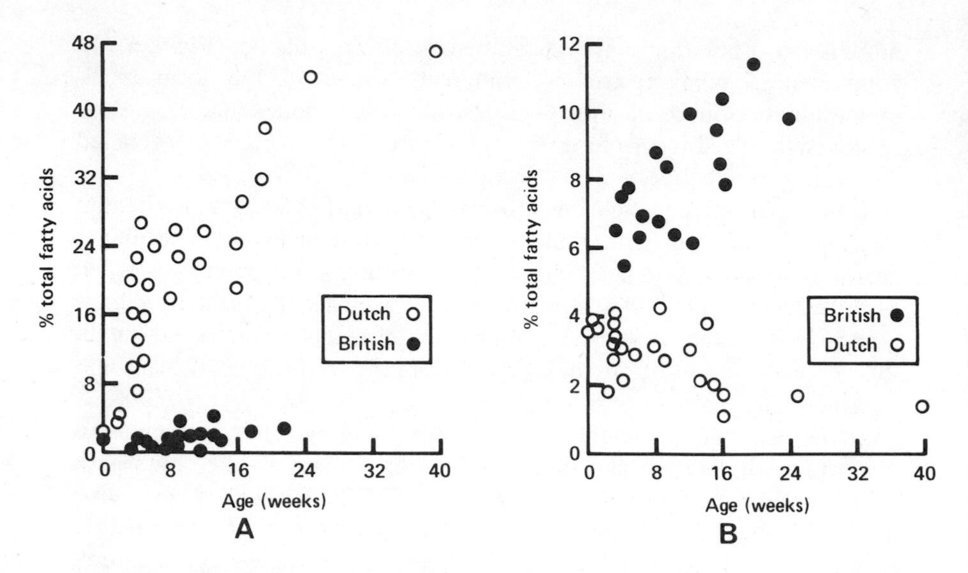

Fig. 5.6. Fatty acids in the adipose tissue of babies fed formulas containing different fatty acids. A, Linoleic acid (*cis*, *cis*-9, 12-octadecadienoic acid); B, myristic acid (tetradecanoic acid). (Adapted from the results of E. M. Widdowson, M. J. Dauncey, D. M. T. Gairdner, J. H. P. Jonxis and M. Pelikan-Filipkova, *British Medical Journal*, 1975, **1**, 653–5.)

5.4. SUMMARY

Fat makes a large contribution to the diet in economically developed countries. It contributes substantially to the palatability of food as well as supplying energy and being a source of essential fatty acids and fat soluble vitamins. Human milk fat is a particularly important source of energy, nutrients and protection against microbes in new-born babies. In supplying artificial formulas, the aim should be to tailor the fat component to the baby's requirements based on modern knowledge. As an aid to reducing the energy content of diets, low fat emulsions and synthetic fats of low digestibility can mimic the properties of conventional food fats, but their efficacy in nutritional and physiological terms is far from certain.

BIBLIOGRAPHY

Burt, R. and Buss, D. H., Dietary fatty acids in the UK, *J. Clinical Practice*, 1984, in press.

Forss, D. A., Odor and flavor compounds from lipids. *Prog. Chem. Fats Other Lipids*, 1972, **13,** 181–258. (A comprehensive account of the role of lipids and lipid breakdown products in the taste, smell and texture of foods. Extensive bibliography.)

Garrow, J. S., *Energy Balance and Obesity in Man*, 1978, Elsevier/North Holland, Amsterdam. (For the reader who wishes to go into more detail on energy metabolism.)

Gurr, M. I., Review of the progress of dairy science: Human and artificial milks for infant feeding, *J. Dairy Research*, 1981, **48,** 519–54. (Gives information on milk composition and the role of milk in the growth and development of the young, the digestion of milk fat and the anti-infective properties of milk fatty acids.)

Kleiber, M., *The Fire of Life: An Introduction to Animal Energetics*, 1961, John Wiley and Sons, New York. (A classic. Fascinating reading for those interested in energy metabolism.)

Ministry of Agriculture, Fisheries and Food, Household food consumption and expenditure, *Annual Report of the National Food Survey Committee*, 1981, HMSO, London. (The world's most comprehensive statistics on a nation's eating habits. Detailed breakdown of amounts and types of fats (and other food components) eaten in Britain.)

Nutrition Society, Symposium on milk consumption and its manipulation, *Proc. Nutr. Soc.*, 1983, **42,** 361–435. (A readable collection of papers including the role of milk in infant feeding, the role of milk in adult diets, manipulation of milk fat composition.)

Rothwell, N. J. and Stock, M. J., *Obesity and Leanness: Basic Aspects*, 1982, J. Libbey and Co., London.

Chapter 6

How Fats are Processed in the Body: Some Basic Metabolism

6.1. INTRODUCTION

It is inappropriate here to provide a detailed account of the metabolism of fats. The interested reader can consult the Bibliography at the end of this chapter. Nevertheless, nutrition is concerned not only with the study of the types and quantities of nutrients required for growth and the health of the whole body, but also with the nourishment of individual living cells. Only by understanding some of the biochemistry involved in the assimilation of individual fat molecules can their role in nutrition truly be appreciated. Moreover, now that the detailed mechanisms of some nutritionally related degenerative diseases are beginning to be understood, it is becoming clear that aberrations in lipid metabolism play a vital role. Some of the material in this chapter, therefore, prepares the ground for the later chapter on fats in health and disease (Chapter 8).

6.2. DIGESTION AND ABSORPTION OF FATS (Fig. 6.1)

After ingestion of food, the first process, occurring in the stomach, is the formation of an oil-in-water emulsion brought about by mechanical movements of churning produced by gastric motility. Lipoproteins are broken down by proteolysis liberating the lipids. Little or no lipolysis (the term used for the enzymic release of fatty acids from their ester linkages – see Fig. 6.2) occurs in the stomach of adult animals. Babies secrete an enzyme, 'lingual lipase', from glands around the tongue that is carried to the stomach and there catalyses the hydrolysis of fatty acids from milk triacylglycerols (see also Section 5.3).

In adults, the process of fat digestion begins in the first part of the

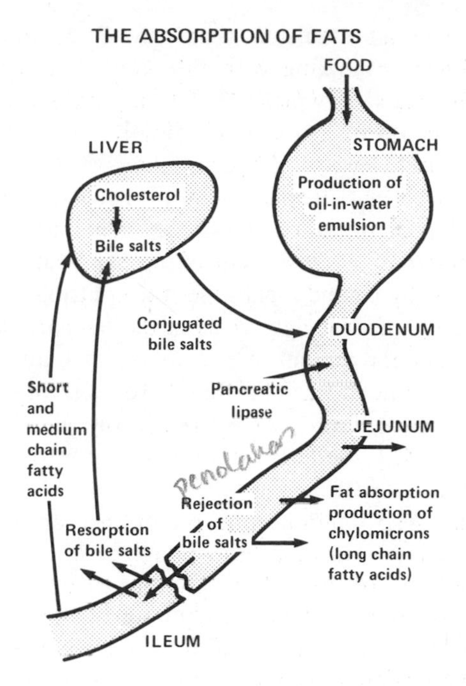

Fig. 6.1. Processes involved in the digestion and absorption of fats in the small intestine. (Reproduced from M. I. Gurr and A. T. James, *Lipid Biochemistry: An Introduction*, 1980, Chapman and Hall, London.)

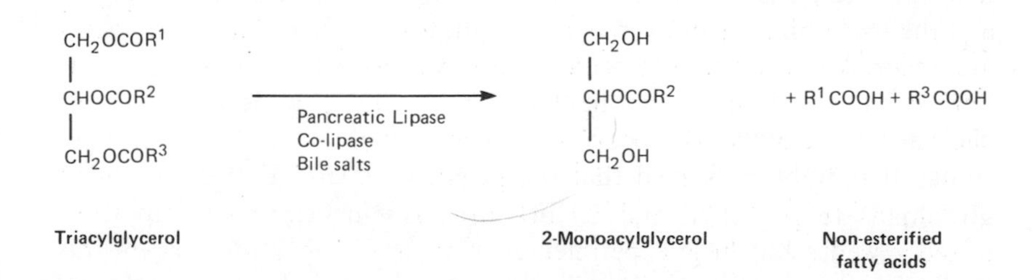

Fig. 6.2. Digestion (lipolysis) of triacylglycerols in the small intestine. The enzyme pancreatic lipase catalyses the breaking of ester bonds in positions 1 and 3 of the triacylglycerols, releasing 2 moles of fatty acid per mole of triacylglycerol; with 1 mole of 2-monoacylglycerol remaining. Bile salts aid in the emulsification of the fat droplets, while colipase anchors the enzyme to the surface of the fat droplets.

small intestine. The fat emulsion that enters the intestine from the stomach is modified by mixing with bile and pancreatic juice. Biliary secretions contain bile salts which are flat amphiphilic molecules with a hydrophobic surface on one side which dissolves in the oil phase at the surface of the fat particles and a hydrophilic surface that interacts with water. In this way the fat droplets acquire a negative electrical charge which attracts a protein called 'co-lipase'. This in turn attracts the enzyme pancreatic lipase a constituent of the pancreatic juice. This is the main fat digesting enzyme and it catalyses the splitting of fatty acids from positions 1 and 3 of the triacylglycerols in the fat particles. The purpose of co-lipase is to anchor the enzyme to the surface and prevent it from being dislodged by bile salts. The main products of fat digestion are therefore non-esterified fatty acids from positions 1 and 3 and 2-monoacylglycerols (with the fatty acid remaining at position 2) (see Fig. 6.2).

Pancreatic juice also contains phospholipases, enzymes that release one of the fatty acids from diacylphosphoglycerides to form lysophospholipids (see legend to Fig. 2.1). Another pancreatic enzyme also converts dietary cholesterol esters into cholesterol and non-esterified fatty acids. The overall process of fat digestion is one which converts fats into more polar derivatives that are more able to interact with water. Indeed the 2-monoacylglycerols, lysophospholipids and the fatty acid soaps are quite strong detergents. As digestion proceeds, monoacylglycerols and non-esterified fatty acids leave the surface of the fat particles and become incorporated into 'micelles', large molecular aggregates of bile salts, lysophospholipids, long chain fatty acids and monoacylglycerols with the hydrophobic parts of the molecules towards the centre of the aggregate and the hydrophilic groups in contact with water. Short chain fatty acids are absorbed by a different route. The mixed micelles are able to draw into the hydrophobic core the less water soluble molecules such as cholesterol, carotenoids, tocopherols and some undigested triacylglycerols. It may be presumed that the products of digestion of the plant glycolipids (Figs. 2.1(h) and 2.1(m)) are also incorporated into these mixed micelles but little experimental work has been published on this topic.

Fat digestion products are absorbed from the mixed micelles across the so called 'brush border' membrane of the enterocytes — the primary absorbing cells in the small intestine — but the details of the mechanism by which they traverse this membrane are poorly understood. Once inside the enterocyte, several events occur. Fatty acids are probably bound to a protein which has the effect of maintaining the flow of fats into the cell. The apparently greater affinity of this 'fatty acid binding' protein for oleic than for stearic acid may in part explain the different rates of absorption of these fatty acids. The flow of fats is also maintained by the continual reconversion of 2-monoacylglycerols to triacylglycerols in the cell. The fatty acids are first converted into an 'activated' form as esters with 'co-enzyme A' (this compound is derived from the B vitamin pantothenic acid). In this form the fatty acids are esterified to the free hydroxyl group of monoacylglycerols in reactions catalysed by separate enzymes. The first step produces diacylglycerol, which is a precursor for phospholipid biosynthesis as well as for triacylglycerol biosynthesis. Active phospholipid biosynthesis is important at this stage because the reformed triacylglycerol has to be 'packaged' in a stable physical form so as to exist in the aqueous environment it now meets on its way to the bloodstream and other body tissues. The particles responsible for this packaging are called 'chylomicrons'. They are predominantly composed of triacylglycerols stabilized by an outer coat of protein, 'apoprotein', and phospholipid. The major phospholipid is phosphatidylcholine, some of which is synthesized within the enterocyte from diacylglycerol. However, some phosphatidylcholine is made by re-esterification of the lysophosphatidylcholine formed from dietary and biliary phosphatidyl choline during fat digestion.

Cholesterol is transported through the intestine much more slowly than triacylglycerols and phospholipids and is slowly incorporated into chylomicrons. Its absorption from the gut lumen is incomplete and about half of what is absorbed is lost by the sloughing off of enterocytes. The bile salts are not absorbed in this part of the small intestine but pass down to the distal part where they are absorbed and recirculated in the portal vein to the liver and thence to the bile where they are available for re-secretion into the gut. This is generally called 'entero-hepatic' circulation and is important in terms of the removal of cholesterol from the body as will be discussed in Chapter 8.

Chylomicrons present an enormous range of particle sizes. Large chylomicrons are produced at the peak of absorption, especially when there is a large proportion of fat in the diet and when the fat is relatively unsaturated. They pass out of the enterocytes into a series of ducts and finally into the blood circulation where they are carried to sites in the body which either store or utilize triacylglycerols.

Chylomicrons, irrespective of the fat eaten, consist mainly of long chain triacylglycerols having fatty acids longer than 14 carbon atoms.

Short chain fatty acids (from dairy products) and medium chain fatty acids (from milk, coconut and palm kernel oil) are selectively transported into the portal blood as non-esterified fatty acids, mainly because of their physical properties. For this reason, medium chain triacylglycerols (MCT) which are refined from coconut oil, are used therapeutically in diets for patients who are unable to absorb long chain fatty acids.

The human gut contains a large number of microorganisms that are capable of modifying fats. Most of these are in the large intestine beyond the sites of fat absorption and fats that are not absorbed are modified in the large bowel, so that the composition of faecal fat is quite different from that in the diet. Insufficient microorganisms are normally present in the gut before the sites of fat absorption make modifications of any nutritional significance, but in gastrointestinal disturbances that result in overgrowth of pathogenic organisms, significant changes can occur. The proliferation of bacteria results in damage to the gut which impairs fat absorption, and increases faecal excretion (steatorrhea). A characteristic excretion product is 10-hydroxystearic acid formed by bacterial modification of dietary stearic acid.

6.3. TRANSPORT OF ABSORBED FATS

The chylomicrons formed in intestinal epithelial cells during the intracellular phase of fat absorption belong to a class of stabilized lipid particles involved in lipid transport called 'plasma lipoproteins' (Fig. 6.3). There are several types which differ in their fat content and they are usually classified according to their density. Thus the chylomicrons, consisting largely of triacylglycerols (about 85%) and only 2% of protein, are the largest and least dense of the lipoproteins. Because of their size, their presence in plasma shortly after a fatty meal can easily be recognised by the opalescence they cause. They are the main carriers of fat derived from the diet. Very low density lipoproteins (VLDL) form a class smaller in size, with less triacylglycerol (about 50%) and about 20% each of cholesterol and phospholipid. They are the main carriers of triacylglycerols formed in the liver from carbohydrates. The next class, yet smaller in size, have a protein content of 50% and about 20% each of phospholipids and cholesterol. Their role is to transport excess cholesterol from membranes to the liver where it can be degraded or converted into bile acids. It should be emphasized that the lipoproteins do not fit

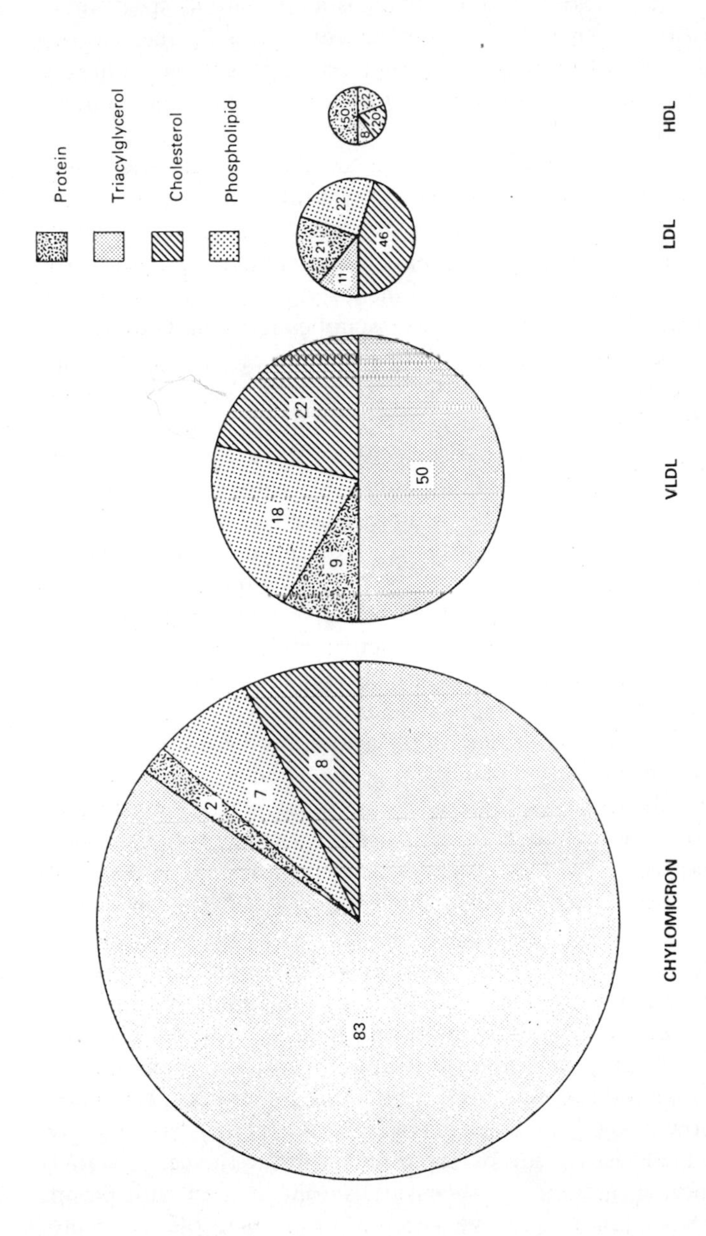

Fig. 6.3. The chemical composition and size ranges of the different classes of human serum lipoproteins. As the particle size decreases, the ratio of protein to fat increases and the density also increases. Chylomicrons are the carriers of dietary fat absorbed and are resynthesized in the intestine. Very low density lipoproteins (VLDL) originate mainly in the liver and carry fats synthesized there from carbohydrates. Low density lipoproteins (LDL) are the main carriers of cholesterol to peripheral tissues. High density lipoproteins (HDL) carry cholesterol to the liver for degradation. (Reproduced from M. I. Gurr and A. T. James, *Lipid Biochemistry: An Introduction*, 1980, Chapman and Hall, London.)

absolutely tidily into these categories. There is a continuous spectrum of particle sizes in the plasma, but it is convenient to classify them in this way, based on methods for separating them from plasma according to their density and such a classification coincides well with their known functions or origins.

Chylomicrons disappear rapidly from the bloodstream and after about 10 min only half the original concentration remains. This occurs mainly by the splitting of fatty acids from the triacylglycerols by the enzyme 'lipoprotein lipase', which is released into blood capillaries close to the adipose tissue or muscle. The activity of the enzyme is controlled by diet and by hormones. After a meal containing significant amounts of fat and carbohydrate, when the supply of energy may exceed the body's immediate needs, the secretion of insulin ensures that the adipose tissue enzyme is active. Fatty acids are released from chylomicron triacylglycerols and taken into adipose tissue where they are re-incorporated into triacylglycerols. The uptake of fatty acids may also be accompanied by the uptake of steroids and fat soluble vitamins. A congenital deficiency of lipoprotein lipase can lead to a condition known as type I hyperlipoproteinaemia characterised by high circulating concentrations of chylomicrons (see also Section 8.3.2). During a fast or in starvation, the adipose tissue enzyme is inactive while the muscle enzyme is 'switched on' and fatty acids are released from circulating lipoproteins, mainly of the very low density type, and taken up into muscle tissue to be used as fuel. Also during fasting, fatty acids may be mobilized from adipose tissue by the activation of a lipase in the fat cells (activated in the fasting state by the hormone adrenalin, inhibited in the fed state by insulin). The fatty acids combine in the plasma with the protein albumin and are transported to the liver where they are synthesized into triacylglycerols to be released into the plasma again, mainly as very low density lipoproteins.

The chylomicrons are not totally degraded by lipoprotein lipase. A smaller, denser, lipoprotein particle, called a 'chylomicron remnant' remains and is further metabolised in the liver to low density lipoproteins which are depleted in triacylglycerol and enriched in cholesterol. These particles provide a supply of cholesterol to tissues. Our understanding of how cholesterol is delivered to the tissues has been revolutionized in the last decade by the discovery that the cells of many tissues have specific receptors on their surface that interact with plasma lipoproteins. After binding to the receptor, the lipoprotein–receptor complex is taken into the cell and the lipoprotein is broken down by

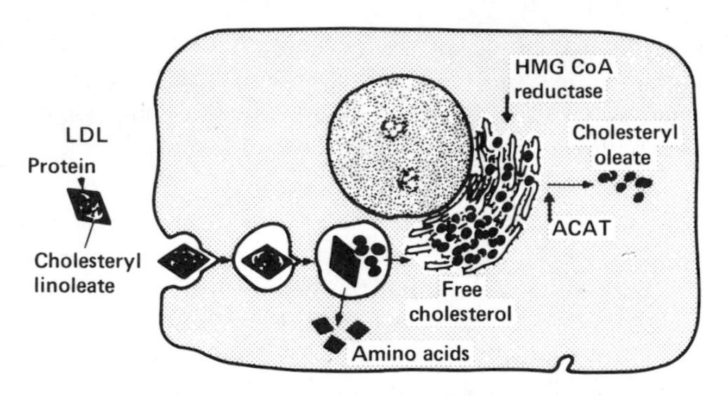

Fig. 6.4. Control of cholesterol metabolism in cells by the lipoprotein receptor mechanism. The cells of many tissues have special proteins on their surface membranes called low density lipoprotein (LDL) receptors. The LDLs from the serum fit specifically into these receptors by a lock-and-key mechanism. After interaction of LDL and receptor has occurred, a piece of the membrane around the receptor 'pinches off' from the main part of the membrane and is carried into the cell. Once inside, the LDL-membrane particle is attacked by enzymes which results in the breakdown of the cholesterol esters in the LDL and the release of free cholesterol. The cholesterol attaches to membranes on which are located the enzymes that catalyse the complete formation of cholesterol from its component parts within the cell. The most critical stage in this process is catalysed by the enzyme HMG-CoA reductase. Thus, when the concentration of LDL cholesterol in the blood is high (reflecting a large intake of dietary cholesterol), large amounts of LDL cholesterol can be taken into cells to suppress the body's own capacity to synthesize cholesterol. When the concentration of cholesterol in the blood is low, the inhibitory mechanism does not occur and the cells can make as much cholesterol as is needed. The concentration of LDL in the blood is also able to determine how many receptors are present on the membrane surface. (Reproduced from J. L. Goldstein and M. S. Brown, *Science*, 1976, **191**, 150–3, by kind permission of the authors and the American Association for the Advancement of Science. © AAAS, 1976.)

enzymes (Fig. 6.4). The resulting free cholesterol interacts with intracellular membranes to inhibit a key enzyme of cholesterol biosynthesis. The intracellular production of cholesterol is thus reduced and does not become activated again until circulating cholesterol levels (reflecting, in part, the dietary intake of cholesterol) begin to fall. When the concentration of cholesterol in the plasma is low, the number of receptors for low density lipoproteins is low and the rate of cholesterol synthesis in the tissue is high. When the concentration of low density lipoprotein in plasma rises, the number of receptors increases. This elegant mechanism enables the body to maintain its supply of cholesterol within well-controlled limits depending on the supply of cholesterol from the diet. This is necessary because cholesterol is a vital compound yet in large concentrations may have undesirable metabolic effects. Several defects in lipoprotein metabolism leading to abnormal blood lipid concentrations have been correlated with defects in the receptor-mediated uptake mechanism.

When the diet contains little fat, the body synthesizes fatty acids from carbohydrates in a wide range of tissues, but mainly the liver (Fig. 6.5). The presence of fat positively inhibits the tissue enzymes that synthesize fatty acids and it is now known that the polyunsaturated fatty acids, especially linoleic acid, have a more potent inhibitory effect than the saturated or monounsaturated ones, although the mechanism of the inhibition is unknown. The fatty acids synthesized by the liver from dietary carbohydrates are mainly incorporated into the triacylglycerols of very low density lipoproteins and exported into the plasma.

For many years, high density lipoproteins received little research attention but in recent years, it has become clear that they play a vital role as scavengers for cholesterol. The cholesterol is present in plasma lipoproteins largely as cholesterol esters. A plasma enzyme (lecithin–cholesterol acyltransferase, LCAT) converts the free cholesterol in high density lipoproteins into cholesterol ester. The cholesterol depleted lipoproteins thus formed interact with membranes, such as those of red blood cells and pick up free cholesterol from the membrane. They then transport this cholesterol to the liver where it is degraded and the high density lipoproteins are returned to the plasma to continue the cycle.

The transport processes described here have great importance in terms of the diseases of lipid metabolism described in Chapter 8.

6.4. BIOSYNTHESIS OF FATS

In the normal healthy body, lipid biosynthesis is regulated in the face of changing needs and dietary intakes. An understanding of the normal regulation of these processes is important in order to be able to combat many diseases in which defects in lipid metabolism seem to play a role (Chapter 8). Most body tissues contain complex systems of enzymes for the biosynthesis of fatty acids and for their esterification in triacylglycerols, phospholipids and other body fats. The rates of lipid biosynthesis in the body are related to the intake of dietary fats (Fig. 6.5).

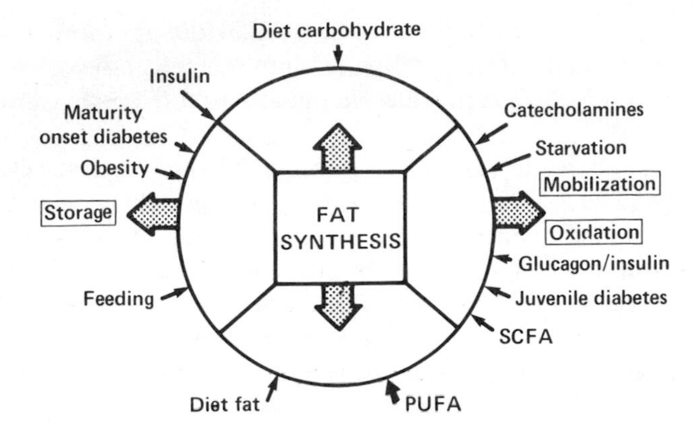

Fig. 6.5. The control of lipid metabolism. The figure illustrates the many factors that may cause fat synthesis (centre) in the body to be *increased* (upper portion) or *decreased* (lower portion) or which may cause fat to be *deposited* (left-hand portion) or *mobilized for oxidation* (right-hand portion).

Dietary carbohydrates and *insulin* tend to increase fat synthesis (upper portion). *Dietary fats*, especially *polyunsaturated fats* (PUFA) suppress fat synthesis (lower portion). During *feeding*, insulin concentrations rise and result in fat *storage* (left-hand portion), which, if excessive, leads to obesity. People with *maturity onset diabetes* in which *insulin* concentrations are high tend to store fat excessively. During fasting, *catecholamines* are released and the ratio of *glucagon to insulin* is high. Under these conditions (right-hand portion) fat *mobilization* is favoured and the released fatty acids are *oxidized* to supply energy. This can occur in *juvenile diabetes* in which *insulin* is lacking. *Short chain fatty acids* (SCFA) in the diet are primarily *oxidized* in the liver not *stored*.

(Reproduced from M. I. Gurr, Dairy fats in human nutrition, in *Fats for the Future*, 1984, Duromark Publishing, Auckland, N.Z.)

When the concentrations of dietary fats are very low, rates of fatty acid synthesis are high, particularly in the liver, in order to supply the needs of the body in respect of structural and storage fats. When the energy intake, as well as the proportion of dietary carbohydrate is high, the rate of fatty acid synthesis is high in the liver, ensuring an ample supply of very low density lipoproteins and the esterification and storage of fat is particularly high in adipose tissue which acts as a reservoir for storage of excess energy. However, this hardly ever occurs in Western man whose diet generally contains a high proportion of energy as fat (about 40% in the UK). This probably means that at most times, the enzymes of fat synthesis are 'switched off' and that the needs for both storage and structural fats are satisfied from dietary intake. A well-balanced metabolic control system must operate so that the appropriate quantities and types of fatty acids are directed into storage, into cellular structures or burned as fuel, even when the intake from the diet is constantly changing.

In general, metabolic control is exercized through alterations in the activity of enzymes in metabolic pathways. Rapid changes may occur when the amount of the enzyme remains constant but its activity is altered by subtle changes in its structure which may occur when the protein combines with small molecules. Longer term adaptive changes are brought about by changes in the absolute number of enzyme molecules without the catalytic activity of individual molecules being affected.

Diet may influence metabolic control in several ways. A very direct way would be the obligatory supply from the diet of a 'coenzyme', a small molecule that is needed to be attached to the enzyme before it can act as a catalyst. The enzymes of fatty acid synthesis require pantothenic acid and biotin as coenzymes. These also happen to be B-group vitamins and must therefore be supplied in the diet. Deficiency of these vitamins therefore leads to defects in lipid biosynthesis which may have profound effects on health. A major role of diet in the control of lipid metabolism is almost certainly to bring about specific changes in the concentrations of circulating hormones which induce or repress the synthesis of some enzymes of lipid metabolism. This is, as yet, a poorly understood area of nutritional biochemistry because of the difficulties of isolating and purifying enzymes that are usually bound very tightly to cellular membranes.

There are many hormones involved in the regulation of lipid biosynthesis and it will be a long while before their mode of action and the complex interactions between them are understood. The dominant role is probably played by insulin. High levels of this hormone characterize the fed state when ample carbohydrate fuel is available from the diet. It suppresses glucose production by the liver and encourages glycogen and fatty acid synthesis. In adipose tissue it encourages the rapid passage of glucose into the cells where it is available for lipid synthesis. It also inhibits the breakdown of fats in adipose tissue. More important, perhaps, than the concentration of a single hormone, is the ratio of the concentrations of different hormones. For example, a high ratio of insulin to glucagon favours esterifications of fatty acids into glycerides and a low rate of fatty acid oxidation. Another characteristic of hormones is their different mode of action in different tissues. Thyroid hormones stimulate the rate of triacylglycerol synthesis in the liver but have the opposite effect in adipose tissue. High concentrations of circulating thyroid hormone are associated with diets rich in carbohydrate. High serum cortisol concentrations result in an elevated rate of fat synthesis in the liver and arise from the ingestion of excessive quantities of saturated fatty acids and sucrose. Obesity is also associated with elevated plasma glucocorticoid concentrations.

Much less is known about the dietary and hormonal control of structural (phospholipid) synthesis because of the difficulties, alluded to above, of studying the controlling enzymes. However, an important factor in controlling membrane lipid synthesis and therefore composition, is the availability of essential fatty acids from the diet and the dietary intake of fatty acids that influence essential fatty acid metabolism. This is discussed in Chapter 7.

6.5. BREAKDOWN OF TISSUE FATS

6.5.1. Hydrolytic Release of Fatty Acids From Glycerolipids

All tissues contain lipases which are enzymes that catalyse the cleavage of ester bonds in lipids. Reference has already been made to adipose tissue lipase and lipoprotein lipase, which are under strict hormonal control. Membranes also contain lipases that cleave fatty acids from different positions of the phospholipid molecules and also the phosphate and base moieties (Fig. 2.1(g)). As a result of the concerted activity of lipases and acyltransferases (enzymes catalysing esterification) there is a continual turnover, or replacement, of all parts of lipid molecules in cells which allows a fine control of their metabolism. Especially important is the membrane phospholipase that liberates essential fatty acids for prostaglandin synthesis (Section 7.2.3.2).

6.5.2. Oxidation of Fatty Acids for Cellular Energy

Oxidation in this context refers to the controlled step-by-step breakdown of fatty acids to yield metabolic energy, not the autoxidation leading to food spoilage discussed in Chapter 3. This is known as β-oxidation and is the chief means by which the energy locked up in fatty acids is made available to the living cell.

β-Oxidation is catalysed by a group of enzymes in a specialized compartment of the cell called the mitochondrion. The complete oxidation of one molecule of the 16-carbon fatty acid palmitic acid yields eight molecules of acetic acid which are further degraded to give carbon dioxide and water. During this process, a substance called 'ATP' is generated. ATP is the most important source of chemical energy in the cell and is required to provide the power for many biosynthetic reactions (including lipid synthesis).

The rate of β-oxidation may be controlled by the supply of precursors or by changes in the activity of the enzymes. These processes are, in turn, subject to dietary and hormonal control. The supply of precursors is determined by the lipolytic release of long chain fatty acids from triacylglycerols, mainly those stored in the adipose tissue, the regulation of which was briefly described in Section 6.3. Diet may also influence β-oxidation by supplying carnitine, a nitrogenous base, which is required for the transport of long chain fatty acids into the mitochondria. Also, two of the enzymes of β-oxidation require a coenzyme, derived from the B group vitamin riboflavin. Deficiency of riboflavin in experimental animals leads to defects in the cellular oxidation of fatty acids.

Although there is some controversy about the relative importance of sugars and fatty acids as fuels in some tissues, it is now accepted that fatty acids are the preferred fuels in most animal tissues. Brain is the prime example of a tissue that cannot utilize long chain fatty acids and this is the main reason why blood glucose levels have to be so strictly maintained in animals. Blood glucose concentration is maintained at a nearly constant level by a series of reactions known as gluconeogenesis, a process which is dependent on the simultaneous β-oxidation of fatty acids.

The flow of fatty acids into pathways which result in their esterification or in their β-oxidation is under hormonal control. A high ratio of glucagon to insulin favours β-oxidation. Insulin suppresses the release of fatty acids from adipose tissue and hence also β-oxidation and gluconeogenesis. In diabetes, where there is a lack of insulin, β-oxidation and gluconeogenesis are enhanced and the products of β-oxidation accumulate faster than the rate at which they can be removed. The result is an excessive accumulation of substances known as ketone bodies. This can occur in any condition where there is a reduced supply of carbohydrates as in starvation or nutritional imbalance.

6.6. SUMMARY

Dietary fats, consisting of a mixture of triacylglycerols, phospholipids, glycolipids, cholesterol, fat-soluble vitamins and pigments, are brought into a coarse emulsion in the stomach and digested by hydrolytic enzymes in the upper small intestine. Bile salts aid the solubilization of digestion products which are then taken up into the intestinal absorbing cells where they are recombined into fats similar to those in the diet. Combination of the absorbed fats with proteins (lipoproteins) enables them to be transported in the bloodstream to tissues where they can be taken up and stored or used as energy according to the body's current needs. The protein part of the lipoprotein carries information which not only specifies the tissues to which the fats should be directed but also influences how they should be metabolized. A large intake of fat signals the tissues to switch off their own production of fats from carbohydrates whereas a high carbohydrate diet signals the tissues to produce their own fat. The chemical substances carrying these signals are hormones of various kinds, the most important being insulin. This complex control system allows the tissues of the body to receive the fats they need and not to overproduce. Defects in these control systems lead to errors in metabolism and metabolic diseases in which nutrition plays a part. These are described in Chapter 8.

BIBLIOGRAPHY

Brown, M. S., Kovanen, P. T. and Goldstein, J. L., Regulation of plasma cholesterol by lipoprotein receptors, *Science*, 1981, **212**, 628. (A readable account of the regulation of lipoprotein metabolism by the cell surface receptor mechanism and its possible implications in disease.)

Brindley, D. N., Digestion, absorption and transport of fats: general principles. In: *Fats in Animal Nutrition*, J. Wiseman (ed), 1984, Butterworths, London.

Gurr, M. I. and James, A. T., *Lipid Biochemistry: An Introduction*, 3rd Edn, 1980, Chapman and Hall, London. (A detailed description of lipid metabolism written mainly for undergraduates. Contains sections on lipid biosynthesis breakdown, digestion and absorption, lipoprotein metabolism, dietary and hormonal control mechanisms.)

Chapter 7

Fats that are Essential in the Diet

7.1. INTRODUCTION

In Chapters 2 and 6, the body's ability to make fats from carbohydrates has been discussed. Most tissues possess the enzymic equipment to convert sugars like glucose into fatty acids via a series of metabolic products of sugar breakdown. They also have the ability to catalyse the esterification of fatty acids with glycerol to make glycerolipids such as triacylglycerols and phospholipids. The most active tissue in this regard is probably the liver. The metabolic machinery for these reactions is only active when there is little fat in the diet, otherwise Nature is idle and uses dietary fats with a minimum of re-structuring. The tissues, however, are unable to make all the fatty acids found in the body or required by the body. The same is true for a variety of other fat-soluble substances not related to fatty acids and usually termed fat-soluble vitamins. These substances have to be supplied in the diet. This chapter describes their role in nutrition and metabolism and what little is known about their quantitative requirements. A distinction has to be made between those fats that are essential in the diet because they are required but cannot be made in the body and those that are essential for proper bodily function yet which can be made by the body tissues, such as cholesterol.

7.2. THE ESSENTIAL FATTY ACIDS

7.2.1. Discovery and Importance of Essential Fatty Acids

The great period of isolation and identification of vitamins was between 1840 and the late 1920s. During that time, the deficiency effects of all the water-soluble and most of the lipid-soluble compounds were demonstrated and by the end of the 1920s it was thought that all the major accessory food factors had been discovered and that carbohydrate and fat were important only in so far as their energy contribution was concerned.

In 1929, Burr and Burr described how acute deficiency states could be produced in rats by feeding fat-free diets and that these deficiencies could be eliminated by adding only certain specific fatty acids to the diet. It was shown that fatty acids related to linoleic acid (see Figs 2.2 and 7.1) were mainly responsible for this effect and the term 'vitamin F' was coined for them, although they are always now referred to as essential fatty acids, usually abbreviated to EFA. EFA deficiency can be produced in a variety of animals, including man, but the condition is best documented in the laboratory rat. The disease is characterized by skin symptoms such as dermatosis and the skin becomes more 'leaky' to water. Growth is retarded, reproduction is impaired and there is degeneration or impairment of function in many organs of the body (Table 7.1). Biochemically, EFA deficiency is characterized by changes in the fatty acid composition of many tissues, especially the biological membranes, whose function is impaired and the production of metabolic energy by the oxidation of fatty acids is reduced. Well-documented EFA deficiency in man is rare, but was first seen in children fed fat-free diets when they developed skin conditions similar to those produced in rats. The skin abnormalities and other signs of EFA deficiency disappeared when linoleic acid was added to the diet. More recently, it has become apparent that EFA deficiency is secondary to a number of disease conditions, so that part of the disease, at least, is amenable to dietary treatment. These conditions will be described in more detail later in the chapter.

7.2.2. Which Fatty Acids are Essential?

Nutritionists and biochemists then began a long and painstaking search for fatty acids with this kind of biological activity. Which fatty acids were essential? What was important about the structure of the molecules that made one fatty acid essential and another not? It was soon discovered that arachidonic acid, a major fatty acid component of animal structural fats, was far more potent than linoleic acid and the first clue to the importance of chemical structure was uncovered. Linoleic and arachidonic acid belong to the same family of polyunsaturated fatty acids with the last double bond in the sequence between carbon atoms 5 and 6 from the methyl end of the carbon chain, i.e. the end of the molecule furthest

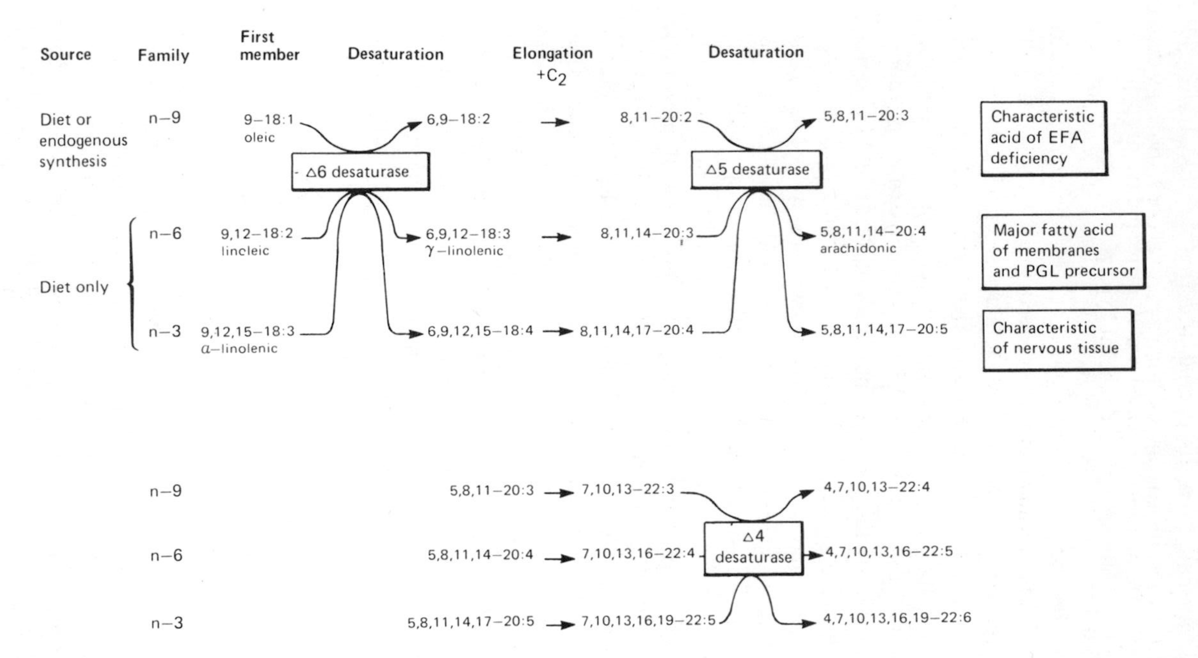

Fig. 7.1. The metabolism of three different families of unsaturated fatty acids. The first member of the *n*-9 family, oleic acid, can be taken in from the diet or can be formed in body tissues whereas linoleic and linolenic acids, the first members of the *n*-6 and *n*-3 families can only come from the diet. The first step in their metabolism is the introduction of a new double bond at position 6. All three fatty acids compete for this enzyme. There follows a series of alternate elongation and desaturation steps. The trienoic acid arising from oleic acid is only formed when linoleic acid is absent from or present in very small quantity in the diet. (Reproduced from M. I. Gurr and A. T. James, *Lipid Biochemistry: An Introduction*, 3rd Edn, 1980, Chapman and Hall, London, by kind permission of the publishers.)

TABLE 7.1
The Major Effects of Essential Fatty Acid Deficiency in Laboratory Rats

Skin	Dermatosis; increased water permeability; drop in sebum secretion, epithelial hyperplasia
Weight	Decrease
Circulation	Heart enlargement; decreased capillary resistance; increased capillary permeability
Kidney	Enlargement, intertubular haemorrhage
Lung	Cholesterol accumulation
Endocrine glands	Adrenals; weight decreased in females and increased in males Thyroid: reduced weight
Reproduction	Males: degeneration of seminiferous tubules Females: irregular oestrus and impaired reproduction and lactation
Metabolism	Changes in fatty acid composition of most organs; increase in cholesterol levels in liver, adrenal and skin Changes in heart and liver mitochondria and uncoupling of oxidative phosphorylation Increased triacylglycerol synthesis and release by liver

Reproduced from M. I. Gurr and A. J. James, *Lipid Biochemistry: An Introduction*, 1980, Chapman and Hall, London.

from the carboxyl group (see Chapter 2, especially Fig. 2.2, and Fig. 7.1).

It was soon discovered that most if not all fatty acids with the *n*-6 structure had essential fatty acid activity. The major fatty acid in green leaves, α-linolenic acid (Figs 2.2 and 7.1), however, also had some activity, though less than linoleic acid. For example, rats fed linoleic acid

Naming of fatty acid families. In the older literature families of fatty acids, identified by the position of the last double bond, were designated by placing the Greek letter ω before the position of the last double bond. Thus linoleic and arachidonic acids belong to the 'ω6' family. Because the official biochemical naming system for fatty acids always numbers fatty acids from the carboxyl group, e.g. linoleic acid is 9,12-octadecadenoic acid (2 double bonds, 9 and 12 carbons from the carboxyl group), the preferred naming of these families is *n*-*x* where *n* is the total number of carbon atoms in the chain and *x* the number of carbon atoms of the last double bond from the end of the chain; thus linoleic and arachidonic acids are '*n*-6' fatty acids in the new naming system.

as the sole source of EFA for two generations grew well and were able to reproduce and lactate successfully. By contrast, linolenic acid could cure only some of the signs of EFA deficiency and when fed as the sole source of EFA, parturition was impaired although many other functions were normal. This acid belongs to another family with the last double bond 3 carbon atoms from the end of the chain ($\omega 3$ or n-3 family). Oleic acid and palmitoleic acid (Figs 2.2 and 7.1) belong to the n-9 ($\omega 9$) and n-7 ($\omega 7$) families respectively and were found to have no essential fatty acid activity. The way in which these fatty acids are related and the reasons for some being essential while others are not, can only be understood by examining the biochemical pathways by which these fatty acids are metabolized in the body and these will be described here only in sufficient detail to enable the reader to appreciate the reasons for fatty acid 'essentiality'.

Animal tissues contain enzymes, called 'desaturases', that insert double bonds into saturated fatty acids, normally at position 9. Hence palmitic and stearic acids (arising either from the diet or by synthesis in the tissues—see Chapter 6) are 'desaturated' to palmitoleic (*cis*-9-hexadecenoic) and oleic (*cis*-9-octadecenoic) acids, respectively. These represent the simplest members of the n-7 and n-9 families referred to above. The enzyme carrying out this transformation is known as a '$\Delta 9$' desaturase. Enzymes are also present that catalyse the insertion of more double bonds to produce polyunsaturated fatty acids, but in animal tissues this only occurs at positions between the first double bond and the carboxyl group. The enzyme responsible for this second desaturation is the $\Delta 6$ desaturase as it inserts a double bond at position 6. Thus oleic acid (n-9) is converted into *cis, cis* -6,9-octadecadienoic acid (n-9) while linoleic acid (n-6) is converted into *all-cis*-6,9,12-octadecatrienoic acid (n-6) and α-linolenic acid (n-3) is converted into *all-cis*-6, 9, 12, 15-octadecatetraenoic acid (n-3) (see Fig. 7.1). It should be noted that in all these metabolically and nutritionally important polyunsaturated fatty acids, the double bonds are always arranged in a particular sequence, separated by a single CH_2 group thus:

$$\cdots -CH_2-CH=CH-CH_2-CH=CH-CH_2\cdots$$

and this 'methylene-interrupted' sequence seems to be absolutely vital to the role of fatty acids in biological tissues. Once a double bond has been inserted at position 6, there is no more room to insert another towards the carboxyl group and still maintain the methylene-interrupted sequence (apparently *cis* double bonds at position 3 are not allowed). Therefore the biochemical strategy is to increase the carbon chain at the carboxyl end by 2 carbon units to create more room for the insertion of another double bond. This elongation process is catalysed by another group of enzymes present in most animal tissue. The scene is then set to introduce another double bond at position 5 (Δ5 desaturase) followed by another elongation and a desaturation at position 4 (Δ4 desaturase) (see Fig. 7.1).

This sequence of desaturations and elongations is the animal's way of producing a variety of polyunsaturated fatty acids to suit its needs and can occur in all four families, *n*-3, *n*-6, *n*-7 and *n*-9. Because, during the course of evolution, animals have lost the ability (still retained by plants) to insert double bonds at positions 12 and 15, *the members of these four families cannot be interconverted in animal tissue.*

Linoleic acid and its relatives are termed 'essential' because without them animals will die. It appears that fatty acids with this structure are vital for reasons explained in Section 6.2.3 and therefore the first member of the family has to be supplied in the diet from plant sources. It can be seen from Fig. 7.1 that arachidonic acid is a product of the elongation and desaturation sequence in animal tissues starting from the dietary precursor linoleic acid. It is therefore an 'essential' fatty acid by the above definition (without it animals cannot function) but it is not essential in the diet as long as linoleic acid is supplied (i.e. it is an 'essential metabolite' but not an essential nutrient). Nevertheless, according to the principles enunciated in earlier chapters, fatty acids supplied from the diet will be incorporated into body fats and therefore dietary arachidonic acid will bypass the sequence of metabolic reactions from linoleic acid to arachidonic acid in the tissues. The fact that the Δ6 desaturase can introduce its double bond at position 6 into the first member of each fatty acid family, *n*-3, *n*-6, *n*-7 and *n*-9, has extremely important consequences. It means that all these fatty acids can compete for the same Δ6 desaturase enzyme and can therefore influence each other's metabolism. Nature has arranged things so that the enzyme does not work on each fatty acid with the same efficiency. The more double bonds present in the starting fatty acid, the more efficiently it is desaturated by the enzyme. Thus the order of affinity of the enzymes for competing fatty acids is α-linolenic > linoleic > oleic > palmitoleic. As far as animals (in particular, man) are concerned, the most important metabolic pathway is the one in which dietary linoleic acid is converted into arachidonic acid. Normally, the diet contains sufficient linoleic acid so that, give its high affinity for the Δ6 desaturase enzyme, this vital metabolic pathway is able to continually supply the quantity of arachidonic acid needed by body tissues. There are circumstances when this is not so however. The first is when the *absolute* amount of linoleic acid in the diet is low. This can happen, for example, when hospital patients are on parenteral or enteral feeds low in fat, or when babies are fed artificial formulas that contain little or no linoleic acid, or in malnutrition. The second is when the dietary intake of linoleic acid is adequate but a person is unable to absorb fat (see Section 6.2). The third is in some genetic disorders resulting in the lack of specific desaturase enzymes. The fourth situation is when the diet contains a small quantity of linoleic acid which is swamped by enormous amounts of other fatty acids in the diet (e.g. oleic acid). In this situation, the fatty acids in excess compete successfully with linoleic acid for the Δ6 desaturase and generate a series of polyunsaturated fatty acids which have no essential fatty acid activity and which cannot substitute for the linoleic acid family. This competition effect is illustrated in Fig. 7.1. It is apparent that, as a result of this competition, the alternative metabolic pathway (beginning with oleic acid) generates a 20-carbon fatty acid with the structure *all-cis*-5, 8, 11-eicosatrienoic acid in place of arachidonic acid. This acid is not normally present in tissue in more than minute amounts and its accumulation provides a biochemical diagnosis of the occurrence and extent of EFA deficiency. Experiments with laboratory animals have shown that even very small quantities of linoleic acid are sufficient to protect the animal from EFA deficiency as long as the remainder of the diet is low in fat. However, increasing the intake of a non-essential fat, such as triolein, while maintaining the same low level of linoleic acid can induce frank EFA deficiency. These experiments illustrate the important concept of a proper balance of fats in the diet which will be enlarged upon in Chapter 8. It will also be apparent from this discussion that while all EFA are polyunsaturated, *not all polyunsaturated fatty acids are EFA*. Many nutritionists and biochemists have used the ratio of 5, 8, 11–20:3/5, 8, 11, 14–20:4 (the triene/tetraene ratio) as a biochemical index of essential fatty acid status. For many years, it was held that a ratio greater than 0·4 was diagnostic of EFA deficiency but recent work has suggested that it may be prudent to revise this figure downwards.

7.2.3. What is the Function of Essential Fatty Acids?

7.2.3.1. Role of Essential Fatty Acids in Membranes

Among the results of EFA deficiency are changes in the properties of biological membranes of which polyunsaturated fatty acids are major constituents in the structural lipids. These changes in properties (such as the permeability of the membrane to water and small molecules such as sugars and metal ions) can be correlated with changes in the fatty acid composition of the membrane. For example, the membranes of liver mitochondria of EFA-deficient animals have smaller proportions of linoleic and arachidonic acid and larger proportions of oleic acid and *all-cis*-5, 8, 11-eicosatrienoic acid than those of healthy animals. These mitochondria are less efficient at oxidizing fatty acids (see Chapter 6) and synthesizing the ATP which is a vital source of chemical energy in the cell. These changes at the molecular and cellular level are reflected in the living animal's poorer performance in converting food energy into metabolic energy for growth and in the maintenance of body function.

The organs and tissues that carry out what might be called the more routine and generalized functions of storage (adipose tissue), chemical processing (liver), fuel utilization (muscle) and excretion (kidney) tend to have membranes in which fatty acids of the *n*-6 family predominate and arachidonic acid is the most polyunsaturated. In contrast, nervous tissue and the retina of the eye have a greater proportion of longer chain acids with 5 and 6 double bonds, predominantly of the *n*-3 family (Fig. 2.2). The main precursor for these acids is α-linolenic acid. It has been assumed that fatty acids of the *n*-3 family have a specific role in vision and in brain function and there is evidence to suggest that when the ratio of *n*-3 to *n*-6 fatty acids is modified experimentally in rat tissues some changes do occur in the electroretinograms and in behavioural responses. The consistent failure to produce more clear-cut abnormalities may imply that α-linolenic acid has no specific function. Nevertheless, it is certainly true that the longer chain metabolites formed from the dietary EFA (arachidonic acid is a good example in the *n*-6 family) are difficult to deplete since the tissues seem to retain these fatty acids very tenaciously. It seems prudent therefore to assume a dietary requirement for α-linolenic acid, at least in certain circumstances, for example in infant feeding (see Chapter 5).

7.2.3.2. Role of Essential Fatty Acids as Precursors of Prostaglandin

While research was being undertaken to determine which fatty acids have EFA activity and how this was related to chemical structure, research was also underway in a very different field that would later add a completely new dimension to the EFA story. Two American gynaecologists reported that the human uterus, on contact with fresh human semen was provoked into either strong contraction or relaxation. The

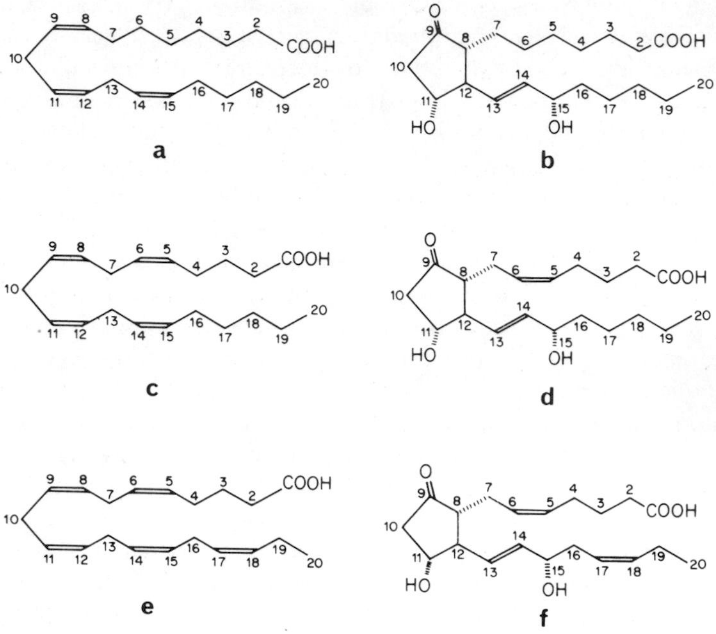

Fig. 7.2. Prostaglandins formed from different polyunsaturated fatty acids. (a) Dihomo-γ-linolenic acid; (b) prostaglandin E_1; (c) arachidonic acid; (d) prostaglandin E_2; (e) eicosapentaenoic acid; (f) prostaglandin E_3.
Prostaglandins of the E-series have a ketone group at position 9 while those of the F-series have a hydroxyl group. Trienoic acids give rise to prostaglandins with 1 double bond (E_1, F_1), while tetra- and pentaenoic acids yield prostaglandins with 2 (E_2, F_2) and 3 (E_3, F_3) double bonds, respectively.

Swede, Von Euler, showed that a fatty acid fraction in lipid extracts from seminal plasma caused marked stimulation of smooth muscle. The active factor was named 'prostaglandin' and was shown to exhibit a variety of physiological and pharmacological properties at extremely low concentrations. When the chemical structures of the prostaglandins were worked out, they suggested to research workers in Sweden and in The Netherlands that the substances might originate from arachidonic acid and this was subsequently proved in a series of very elegant experiments.

The prostaglandins are oxygenated fatty acids with an unusual ring of 5 carbon atoms (Fig. 7.2). Arachidonic acid can generate a whole range of related compounds but with subtle differences in structure (Fig. 7.3), which exert a range of profound physiological activities at concentrations

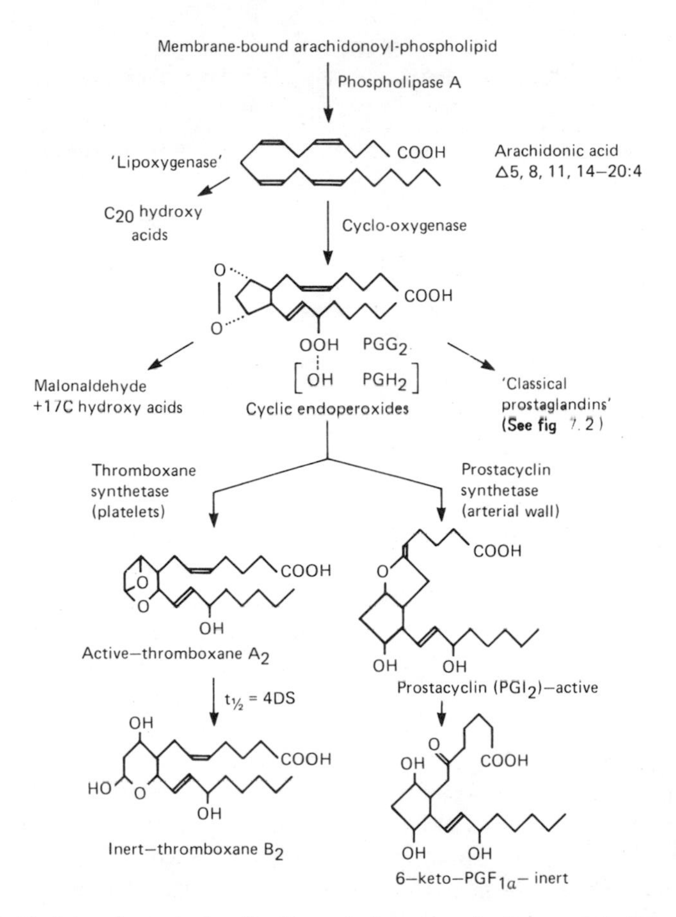

Fig. 7.3. Variety of prostaglandin-like substances and other metabolites formed from arachidonic acid.

down to 10^{-9} g g^{-1} tissue. These include the ability to contract smooth muscle, to inhibit or stimulate the adhesion of blood platelets and to cause constriction or dilation of blood vessels with a related influence on blood pressure.

Two groups of these metabolites, the 'prostacyclins' and the 'thromboxanes', have essentially opposite physiological effects (Table 7.2). Prostacyclins, formed in arterial walls, are among the most powerful known inhibitors of platelet aggregation. They relax the arterial walls and promote a lowering of blood pressure. Thromboxanes, found in

TABLE 7.2
Opposing Effect of Prostacyclins and Thromboxanes on the Cardiovascular System

Thromboxanes in platelets	*Physiological effect*	*Prostacyclins in arterial wall*
Stimulates	Platelet aggregation	Inhibits
Constricts	Arterial wall	Relaxes
Lowers	Platelet cyclic AMP concentration	Raises
Raises	Blood pressure	Lowers

Reproduced from M. I. Gurr and A. T. James, *Lipid Biochemistry: An Introduction*, 1980, Chapman and Hall, London.

platelets, stimulate platelets to aggregate (an important mechanism in wound healing), contract the arterial wall and promote an increase in blood pressure. The balance between these activities is important in maintaining normal vascular function and if thrown into imbalance (which may be subject to dietary influence — see Chapter 8) the progress of vascular disease may be hastened. They are so potent in their action that they need to be generated locally and destroyed immediately, after they have produced their effect, by enzymes that convert them into inactive metabolites. The excretion of these metabolites in urine has been used as a method for estimating daily production of prostaglandins and assessing the quantities required by the body. In order to achieve this local production, the essential fatty acids that are precursors of the prostaglandins are released from membrane phospholipids by specific phospholipases (see Chapter 6) and transferred to the enzyme that synthesizes prostaglandins, which is also located in the membrane. The structural, or membrane, phospholipids can therefore be regarded as a vast body store of essential fatty acids that are immediately available for prostaglandin synthesis when required. As they are used up, they must be replenished by new fatty acids coming in from the diet (or, like arachidonic acid) synthesized from dietary linoleic acid. This explains the need to maintain a dynamic turnover of fatty acids in membranes.

Arachidonic acid is not the only essential fatty acid to give rise to prostaglandins. Its precursor di-homo-γ-linolenic acid (also of the *n*-6 family) and some even more unsaturated fatty acids of the *n*-3 family can be converted into prostaglandins. Fatty acids with 3, 4 and 5 double bonds which give rise to prostaglandins with 1, 2 and 3 double bonds, respectively, are named as the 1, 2 and 3 series of prostaglandins (Fig. 7.2). Indeed, one criterion of essentiality in a fatty acid is that it is capable of being converted into a prostaglandin with physiological activity. The daily production rate of prostaglandins in man has been estimated at about 1 mg day^{-1} and this has to be set against the daily intake of around 10 g or so of essential fatty acids (see Section 7.3). It appears that the membrane functions and prostaglandin function of essential fatty acids cannot be separated entirely but that nevertheless, as discussed in the previous section, the presence of a certain quantity of specific EFA has to do with the integrity of membranes in a way which may have little relation to the provision of prostaglandins.

7.2.4. How Much Essential Fat is Required in the Diet?

A precise answer to this question is impossible at present for a number of reasons. Firstly (as indicated in Sections 7.2.3.1 and 7.2.3.2), we are not entirely sure of all the functions of EFA and the extent to which different EFAs have distinct functions. It seems possible, too, that there are different levels of requirements from EFA for their different functions. For example, there may be a relatively low level of requirement (as for most vitamins) which is necessary to prevent the overt signs of EFA deficiency such as skin lesions and growth retardation. This may be around 1% of dietary energy as linoleic acid. At a higher level, there may be a grosser but quantitatively less easily definable quantity required for such functions as maintaining an optimal balance of membrane fatty acids, preserving a reservoir for prostaglandin formation and maintaining optimal cholesterol levels (see also Chapter 8). The absolute amounts required at this level are likely to be influenced by the amounts and types of other fats in the diet because of the competition for metabolic pathways between EFA and non-EFA described in Section 7.2.2.

Secondly, experimental data are in general available only for laboratory animals such as rats and mice. The indications are that different species have different requirements, both quantitative and qualitative and, therefore, man's requirements can only be implied. A good example of a qualitatively different requirement between species, is that for the cat family; arachidonic acid is not merely an essential metabolite as in other mammals but an essential nutrient because cats do not possess enzymes capable of converting linoleic acid into arachidonic acid.

Thirdly, it seems that requirements change with age and with physiological conditions. Thus the requirements of the foetus differ from those of the neonate which are in turn different from those of adolescents, pregnant and lactating women and the elderly. In human beings, there is no sound way of approaching the problem of requirements experimentally. We must gradually build up a picture from observation of signs and symptoms occurring under conditions of a known intake of EFA.

These considerations clearly indicate that another problem in assessing requirements is in choosing the criteria by which EFA requirements are judged. Are they to be related to the amounts required by animals to cure signs of EFA deficiency? Should a more rigorous biochemical criterion be used such as the ratio of *all-cis*-5, 8, 11–20:3/*all-cis*-5, 8, 11, 14–20:4 in tissue lipids? Dietary linoleic acid at a concentration of 1–2% of the total dietary energy will cure signs of EFA deficiency in rats and reduce the elevated triene/tetraene ratio. Obtaining this kind of evidence in human beings is usually fortuitous. There has been one experiment in which 400 infants were fed milk formulas containing different amounts of linoleic acid. When the formulas contained less than 0·1% of the dietary energy in the form of linoleic acid, chemical signs of EFA deficiency ensued with a characteristic increase in the triene/tetraene ratio in serum lipids. The minimum requirements for linoleic acid were judged to be about 1% of the dietary energy. Although the linoleic acid content of human milk is more than enough to satisfy the requirements of a breast-fed baby, on these criteria, the presence of di-homo-γ-linolenic acid (the immediate precursor of arachidonic acid (see Figs 7.1 and 7.2)) has been suggested as indicating that the efficiency of conversion of linoleic acid into arachidonic acid is not sufficient for the baby's needs.

One of the most detailed assessments of probable human requirements for different EFAs is in a report by FAO/WHO (see Bibliography). In the absence of experimental evidence for precise requirements, they have made estimates for the requirements in pregnancy based on the amounts of EFA known to accumulate in the foetus and in the maternal adipose tissue which amounts to between 600 and 700 g. On this basis the EFA requirements in pregnancy are increased by 1–1·5% of dietary energy above the basal requirements which are usually taken as about 1% of dietary energy. Similarly, lactational requirements can be estimated from the average content of EFA in human milk and the volumes of milk produced. On this basis, the additional EFA requirements of lactating mothers would be increased by 1–2% of dietary energy. Thus, depending on physiological needs, the best estimates that can be made for human EFA requirements are between 1 and 3% of dietary energy.

The results of more recent research are showing that EFA deficiencies, secondary to other diseases, are perhaps more common than we thought.

The necessity to devise diets to cure these deficiencies is throwing more light on quantitative and qualitative human requirements.

'Total parenteral nutrition' has become an important life-saving technique in modern clinical practice. However, early practice was to employ fat-free preparations which soon induced signs of EFA deficiency. Serum concentrations of all members of the *n*-6 family are subnormal and 20:3(*n*-9) can be elevated more than 10-fold. It is interesting that the elevation of 20:3(*n*-9) occurs only when *both* linoleic and linolenic acids are deficient in the diet because the presence of *either* *n*-6 or *n*-3 acids in the diet suppresses the biosynthesis of 20:3(*n*-9). The triene/tetraene ratio is useful, therefore, only in circumstances when linolenic acid is not present in the diet and the modern approach to the assessment of EFA deficiency is to measure all possible isomers of serum polyunsaturated fatty acids by capillary gas-liquid chromatography.

Feeding by total parenteral nutrition can also throw light on the possible requirements for *n*-3 fatty acids in human beings, which have hitherto been difficult if not impossible to assess. In one case, neurological symptoms occurred after 4 or 5 months on parenteral nutrition in which the fat component was a safflower oil emulsion containing mainly linoleic acid and only a minute amount of linolenic acid (*n*-3). When the

TABLE 7.3
Changes in Polyunsaturated Fatty Acids of Serum Phospholipids in Chronic Malnutrition

Fatty acid	*Ratio of serum concentration compared with that in a normal population*[a]
18:2, *n*-6 (linoleic)	0·84
18:3, *n*-6 (γ-linolenic)	0·53
20:2, *n*-6	1·90
20:3, *n*-6	0·78
20:4, *n*-6 (arachidonic)	0·70
22:4, *n*-6	2·70
22:5, *n*-6	0·85
18:3, *n*-3 (α-linolenic)	1·90
n-3, metabolites	1·10

Adapted from R. T. Holman, Essential fatty acids in nutrition and disease, *Chemistry and Industry*, 1981, 704–9.

[a] These fatty acid profiles were measured in a group of Argentinian children suffering from chronic malnutrition.

safflower oil was replaced by soybean oil, containing much more α-linolenic acid, the neurological symptoms disappeared.

Chronic malnutrition in many parts of the world is associated with a low fat intake and consequently a risk of EFA deficiency. In malnourished children, the proportion of linoleic acid and related acids is often very low (see Table 7.3); yet in some cases the proportions of long chain *n*-6 fatty acids can be abnormally high, indicating some metabolic aberrations associated with protein or other nutrient deficiencies.

Several metabolic diseases often with a genetic basis are associated with EFA deficiencies as measured by total fatty acid profiles of serum. Often this deficiency arises because an aspect of the disease is an inability to absorb fat as in cystic fibrosis, achrodermatitis enteropathica and Crohn's disease, sometimes because of a failure to metabolize the EFAs efficiently, as in Sjögren–Larsson syndrome, Reyes syndrome and Prader–Willi syndrome.

7.3. FAT-SOLUBLE VITAMINS

The view that the only components of the diet necessary for health, growth and reproduction were protein, fats, carbohydrate and certain minerals had to be changed when it was realized early this century that minute amounts of additional substances were also required. These 'accessary food factors' appeared to be either fat-soluble or water-soluble and were given the name 'vitamins' and divided into fat-soluble A and water-soluble B, respectively. Eventually, it was realized that both these classes were mixtures of chemically unrelated compounds and the fat-soluble vitamins, found mainly in fatty foods, were later classified as A, D, E, and K.

Requirements for the fat-soluble vitamins are of a different order of magnitude from the essential fatty acids being measured in micrograms per day rather than grams per day.

7.3.1. Vitamin A

The chemical name for vitamin A is retinol (Fig. 7.4), although retinol itself is not the only source of vitamin A activity in foods. Retinol is only found as such in animal fats. Vegetables, such as dark green leaves and vegetable oils like palm oil contain a precursor (provitamin A), β-carotene, which is converted into retinol in the body. Because there are losses during the absorption of carotene and its conversion into retinol, it

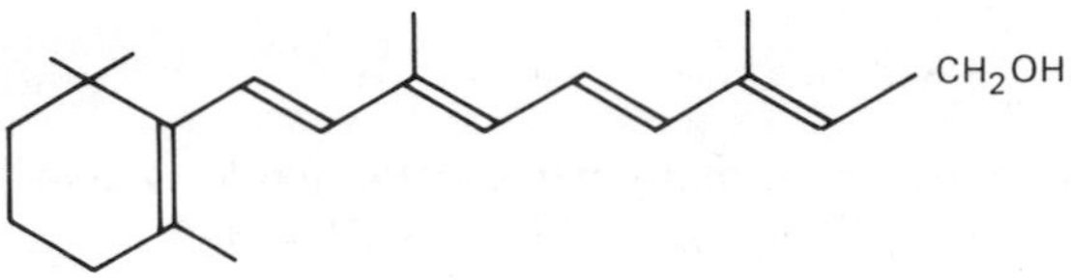

CH_2OH

Fig. 7.4. Retinol (vitamin A).

is convenient to describe the vitamin A activity of foods in terms of 'retinol equivalents'. Normally, 1μg retinol equivalent is taken as corresponding to 6μg β-carotene, although there are strong indications that this factor generally underestimates the retinol equivalents of most diets. Milk fat is relatively rich in retinol equivalents and because of the better absorption of β-carotene from milk fat than from many other foods, 1μg retinol equivalents corresponds to only 2μg β-carotene in milk. The carotene content of milk depends on the diet of the cow (higher in cows on pasture than on forage), the season (higher in spring and early summer than in winter) and breed (higher in Channel Island than in Friesian cattle because of the lower conversion into retinol in the gut). This accounts for the yellower colour of Channel Island compared with Friesian milk. The concentration of retinol in dairy foods is proportional to the concentration of milk fat in the product. Vitamin A is not widely distributed in foods (Table 7.4). The main sources are green vegetables, carrots, liver, milk, butter and margarine. In the UK, all margarine for retail sale is required by law to contain about the same amount of vitamin A (added as synthetic retinol or β-carotene) as butter. Fish liver oils are by far the most concentrated source, but mammalian storage fats such as lard and dripping contain none. The British diet provides on average about twice the recommended intake of vitamin A, with two-thirds coming from retinol itself and about a third from carotene. This happy situation is not shared by many parts of the world, since vitamin A deficiency is widespread, the most afflicted countries being Bangladesh, India, Indonesia and The Philippines. The problem also exists, though to a lesser degree, in many African and Central or South American countries. Vitamin A is essential for vision and the most tragic manifestation of vitamin A deficiency is blindness in young children. The first effects are seen as severe eye lesions, a condition known as xerophthalmia which is eventually followed by keratomalacia with dense scarring of the cornea and complete blindness. Xerophthalmia is considered to be one of the four commonest preventable diseases in the world. Although there are

TABLE 7.4
The Vitamin A and β-Carotene Content of some Raw Foods ($\mu g\ 100\ g^{-1}$)

Food	*Retinol*	*β-Carotene*	*Retinol equivalent*
Animal foods			
Milk			40
Cheese (Cheddar)			410
Eggs	140		
Beef			
Liver (Ox)			16 760
Kidney	120		
Cod (flesh)			
Herring (flesh)	45		
Sardines (canned)			
Butter			985
Margarine			900
Cod liver oil	18 000		
Halibut liver oil	900 000		
Vegetable foods			
Potatoes		0	0
Cabbage		300	50
Spinach		6 000	1 000
Peas (fresh or frozen)		300	50
Watercress		3 000	500
Carrots (old)		12 000	2 000
Tomatoes		600	100
Apricots (dried)		3 600	600
Flour			

Adapted from D. H. Buss and J. Robertson, *Manual of Nutrition*, 1978, Ministry of Agriculture, Fisheries and Food, HMSO, London.

large scale programmes for the supplementation of children's diets with vitamin A, these are difficult to implement successfully and the World Health Organization considers that if the consumption of green leafy vegetables and suitable fresh fruits by young children could be substantially increased, there is every reason to believe the problem would be solved. There might also be a case for trying to increase the overall fat content of the diet, too. Here we are more concerned with solving problems of economics, distribution and with changing local eating habits than with nutrition, where the knowledge is already to hand. Although 'developed' nations do not share the problems of xerophthalmia, epidemiological evidence shows that people with above average blood retinol concentrations or above average β-carotene intakes have a lower average risk of cancer. The relationship between vitamin A and cancer is the subject of much current research and may be related to the involvement of vitamin A in maintaining the normal differentiation of specialized epithelial tissues throughout the body (i.e. those tissues that are mainly involved in the processes of absorption or secretion). Deficiency of vitamin A results in failure of differentiation and therefore failure of normal functions. The exact mechanisms, however, are obscure.

There is a mistaken idea amongst many laymen that because vitamins are essential for health, the more that are consumed the more effective they will be. This is quite fallacious and most vitamins are required in exceedingly small quantities. (Thus, no more than $\frac{3}{4}$ mg (1/36000 ounces!) of vitamin A are required by the average person per day.) Whereas an excess of most water-soluble vitamins is dealt with by the body by excretion, this is not so with fat-soluble vitamins. Vitamin A, if taken in excess, will accumulate in the liver and eventually destroy it. Xavier Mertz, the Antarctic explorer, was forced, during an expedition in 1912 to eat the raw livers of his dogs for sustenance. He died from the effects of excessive vitamin A which resulted in the complete peeling of the skin from the whole of his body. Such can be the effects of mega-vitamin therapy!

7.3.2. Vitamin D

Vitamin D is a general name for a family of steroids with anti-rachitic

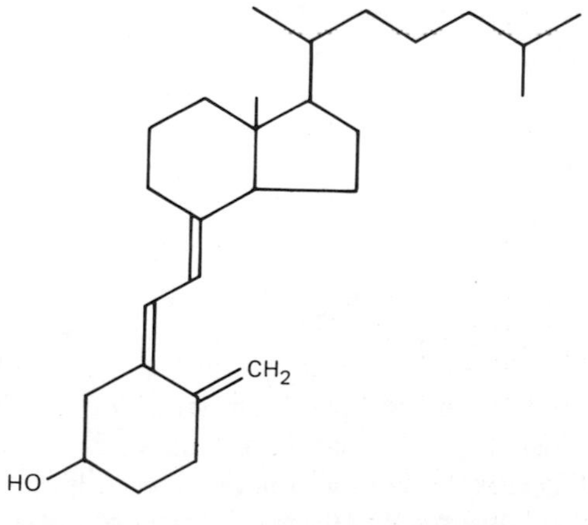

CH2
HO

Fig. 7.5. Cholecalciferol (vitamin D).

TABLE 7.5
Vitamin D Content of some Raw Foods
(μg 100 g^{-1})

Milk (liquid, winter)	0·01
Milk (liquid, summer)	0·03
Milk (UHT)	0·02
Milk (evaporated)[a]	2·91
Cheese (Cheddar)	0·26
Eggs	1·75
Liver	0·75
Herring, kipper	22·40
Salmon (canned)	12·50
Sardines (canned)	7·50
Butter	0·76
Margarine[a]	7·94
Ovaltine (dry)[a]	30·60
Cod liver oil	212·50

Adapted from D. H. Buss and J. Robertson, *Manual of Nutrition*, 1978, MAFF, HMSO, London.
[a] Includes added vitamin D.

properties. The only one of significance is the one now called vitamin D_3 or cholecalciferol (Fig. 7.5). It is present in only a few foods (Table 7.5), the richest source being the liver oils of fish.

Cholecalciferol is produced by the ultraviolet irradiation of 7-dehydrocholesterol, a sterol widely distributed in animal fats including the skin surface lipids.

Many people get little or no vitamin D from their diet but obtain all they require from the action of the ultraviolet rays in sunlight on the 7-dehydrocholesterol in the skin. For this reason, many nutritionists have argued that vitamin D should hardly be considered a vitamin at all. However, two groups of people may have a special need to obtain vitamin D from the diet. In the first group are children and pregnant and lactating women whose requirements are particularly high. In the second group are people who are little exposed to sunlight such as the housebound elderly and people in northern latitudes or those who wear enveloping clothes. Dark-skinned immigrants to Northern European countries are especially vulnerable. Infants and children who obtain too little vitamin D develop rickets, with deformed bones that are too weak to support their weight. The reason why, in the UK and some other countries, vitamin D preparations are provided for children and pregnant women and margarine is fortified with it is because these degenerative changes soon become permanent if supplementation is not begun early enough.

The main role of vitamin D is to maintain the concentrations of calcium and phosphorus in the blood, primarily by enhancing the absorption of dietary calcium from the alimentary tract and regulating the interchange of calcium between blood and bone. Like vitamin A, vitamin D is toxic in high doses. Too high an intake causes more calcium to be absorbed than can be excreted, resulting in excessive deposition in, and damage to, the kidneys. Dietary vitamin D is absorbed along with the bulk of the dietary fat and carried into the blood in the chylomicrons. In the liver, it is converted into a substance called 25-hydroxycholecalciferol which is further converted into 1,25-dihydroxycholecalciferol in the kidneys. The latter can be regarded as a hormone that is involved in the regulation of calcium metabolism.

7.3.3. Vitamin E

Because vitamin E is so widespread in foods and, like other fat-soluble vitamins is stored in the body, deficiency states are rarely if ever seen, a possible exception being in premature infants with low fat stores. Indeed, although vitamin E is known to be necessary for normal fertility in rats, it has never been proven to be necessary for man. The activity is shared by a family of compounds known as tocopherols, the most active of which is α-tocopherol (Fig. 7.6). The richest sources are vegetable oils, cereal products and eggs. α-Tocopherol is present in the lipid bilayers of biological membranes (see Chapter 2) and may play a structural role there. It is known to be a powerful antioxidant and prevents the autoxidation of unsaturated lipids described in Section 3.3.7. The products of autoxidation of unsaturated fatty acids can cause damage to

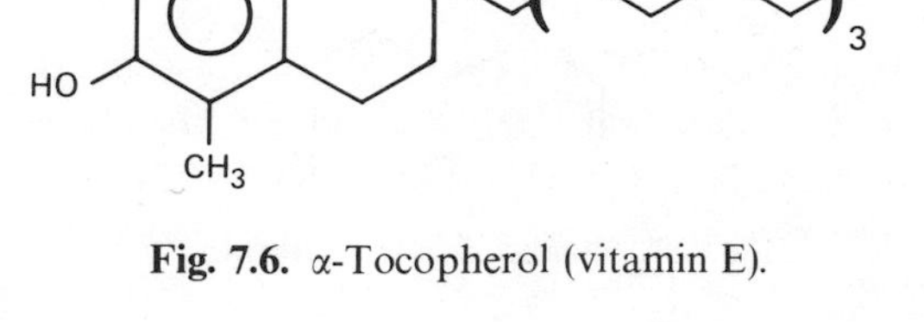

CH_3 H_3C O CH_3 CH_3 H HO 3 CH_3

Fig. 7.6. α-Tocopherol (vitamin E).

cells if the oxidative process is not kept in check and such damage appears to be exacerbated in animals fed diets deficient in vitamin E. These observations have led to the postulate that vitamin E plays a vital role in the prevention of oxidation of membrane fatty acids in living cells, but despite many decades of research, the precise role of vitamin E is still obscure. It is a generally held view that intake should be considered in relation to the polyunsaturated fatty acid content of the diet rather than in absolute amounts. A ratio of vitamin E to linoleic acid of about 0·6 mg g^{-1} is generally recommended. In general, those vegetable oils containing high concentrations of polyunsaturated fatty acids are sufficiently rich in vitamin E to give adequate protection.

7.3.4. Vitamin K

Vitamin K activity is shared by a group of chemical compounds called menaquinones (Fig. 7.7). They are produced mainly in green leafy vegetables or by bacteria. Indeed, the reason why no well-defined human deficiency has been described may be because our intestinal bacteria are capable of providing for our entire needs.

Vitamin K is necessary for the normal clotting of blood and a deficiency would prolong the time taken for blood to clot.

Deficiencies of vitamin K and indeed all the other fat-soluble vitamins are more likely to occur as a result of impairment in fat absorption than from dietary insufficiency. This could occur when the secretion of bile salts is restricted (as in biliary obstruction), when sections of gut have been removed or damaged by surgery or in diseases, such as 'tropical sprue' and cystic fibrosis, that are associated with poor intestinal absorption. Even when normal absorptive mechanisms are functioning well, some fat is necessary in the diet to improve the absorption and utilization of fat-soluble vitamins. However, there is little evidence that,

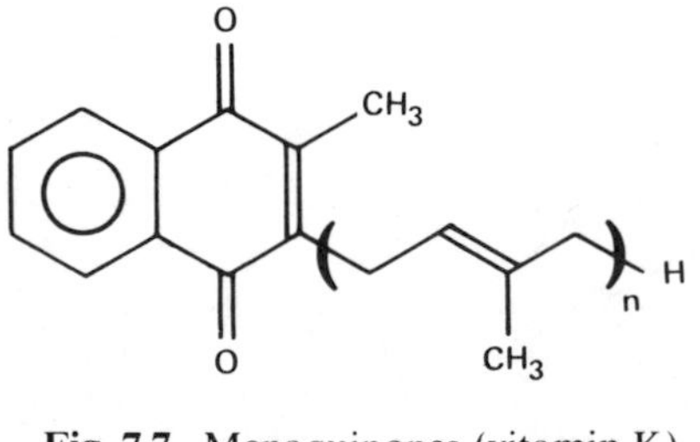

O
CH_3
H
n
O
CH_3

Fig. 7.7. Menaquinones (vitamin K).

within the normal range of fat intakes, the amount of dietary fat significantly affects the utilization of fat-soluble vitamins.

7.4. SUMMARY

Whereas the human body has the capacity to manufacture for itself saturated and monounsaturated fats and cholesterol, it is unable to synthesize linoleic acid and vitamins A, D, E and K. These substances are, however, essential to life and therefore need to be obtained from the diet. The amounts required daily in the diet can be assessed albeit rather imprecisely. Requirements vary between individuals and depend on the rate of growth, period of life and on physiological states such as puberty, pregnancy and lactation.

Linoleic acid is converted into a series of longer chain, more highly unsaturated fatty acids in the human body and these act as precursors for a variety of hormone-like substances with powerful physiological activities (the prostagladins). The metabolism of essential fatty acids is influenced by the level and types of other fats in the diet. The supply of fat-soluble vitamins to body tissues is determined not only by the concentrations in the diet but by factors that influence the efficiency of absorption of fats in general.

BIBLIOGRAPHY

Curtis-Prior, P. B., *Prostaglandins*, 1976, North-Holland Publishing Co., Amsterdam.

Cuthbertson, W. F. J., Essential fatty acid requirements in infancy, *Am. J. Clin. Nutr.*, 1976, **26**, 559–68. (The author points to the rarity of essential fatty acid deficiency and argues that requirements have been overstated — cf. Soderhjelm *et al.* below.)

Food and Agriculture Organization, *Dietary Fats and Oils in Human Nutrition: A Joint FAO/WHO Report*, 1978, FAO, Rome. (One of the few documents that attempts to evaluate quantitative requirements for essential fatty acids in different physiological states.)

Gurr, M. I., The role of lipids in the regulation of the immune system, *Prog. Lipid Res.*, 1983, **22**, 257–87. (Reviews the evidence that lipids, including essential fatty acids and prostaglandins, are involved in the regulation of immune function.)

Holman, R. T., Biological activities of and requirements for polyunsaturated fatty acids, *Prog. Chem. Fats Other Lipids*, 1970, **9**, 607–82. (A detailed account of the structures, activities and biochemistry of the known fatty acids with essential fatty acid activity.)

Holman, R. T. and Johnson, S. B., Essential fatty acid deficiencies in man. In: *Dietary Fats and Health*, E. G. Perkins and W. J. Visek (Eds), 1983, American Oil Chemists Society, Champaign, Illinois.

Lands, W. E. M., The biosynthesis and metabolism of prostaglandins, *Ann. Rev. Physiol.*, 1979, **41**, 633–52. (A detailed review for the reader who wishes to go into the more biochemical aspects of prostaglandins.)

Ministry of Agriculture, Fisheries and Food, *Manual of Nutrition*, eds D. H. Buss and J. Robertson, 1978, HMSO, London. (Gives some basic information on nutritional aspects of fat soluble vitamins.)

Nutrition Society, Symposium: Vitamin A in nutrition and disease, *Proc. Nutr. Soc.*, 1983, **42**, 7–101. (A collection of papers giving an up to date account of research in vitamin A as well as an historical background. For more detailed accounts of vitamins K and E the reader should refer to Chapters 26 and 27 in Perkins and Visek (see Bibliography to Chapter 4).)

Soderhjelm, L., Wiese, H. F. and Holman, R. T., The role of polyunsaturated fatty acids in human nutrition and metabolism, *Prog. Chem. Fats Other Lipids*, 1970, **9**, 555–85. (The authors describe their classic work on essential fatty acid deficiency and argue for substantial requirements — cf. Cuthbertson above.)

Vergroesen, A. J. (Ed.), *The Role of Fats in Human Nutrition*, 1975, Academic Press, London. (Contains chapters on the diverse roles of linoleic acid including its hypolipidaemic effect, influence on platelet aggregation, role in hypertension and influence on vitamin E requirements.)

Chapter 8

Fats in Health and Disease

8.1. INTRODUCTION: THE CONCEPTS OF NUTRIENT REQUIREMENTS AND A 'BALANCED' INTAKE OF NUTRIENTS

The Concise Oxford Dictionary defines health as 'soundness of body' and it is clear that nutrition must play an important role in maintaining sound body functions. This is because some components of the diet are absolutely required to support the body chemistry whilst others provide the fuel for that chemistry to be maintained. Chapters 2, 5, 6 and 7 of this book have explained how the fatty components in the diet fulfil these functions and, to this extent, the discussion so far has largely been about the role of fats in the healthy body. By contrast, disease is defined as 'a morbid or unwholesome condition of the body', and we can infer that inappropriate nutrition is at least one factor in contributing to disease.

Building from these fundamental concepts, it is useful to construct a chart which relates the level of intake of nutrients with bodily function and health (Fig. 8.1). In the complete absence of a nutrient that is a necessary component of the diet, essential body chemistry cannot function and life cannot long be sustained as represented by the far left of the diagram. Such circumstances are rare except in conditions of complete starvation, although it could conceivably occur when a person insists on taking or is forced by circumstances to take a very extreme type of diet, or when the essential nutrient has been eliminated by processing or poor storage. More common would be the situation in which a diet provides a small but inadequate intake of the nutrient, leading to overt deficiency signs. Good examples in the realm of fat nutrition would be deficiency diseases occurring as a result of inadequate intakes of essential fatty acids, vitamin A or vitamin D. Table 8.1 gives a

TABLE 8.1
Human Deficiency Diseases Associated with Essential Fats

Vitamin A	Night blindness; eye lesions (xerophthalmia); complete blindness (keratomalacia)
Vitamin D	Bone softening (osteomalacia) in adolescents, pregnant women and elderly Bone softening (osteomalacia) in adolescents, pregnant women and elderly
Linoleic acid (*n*-6)	Dermatosis; increased water permeability of skin; decreased sebum secretion; epithelial hyperplasia; weight loss; changes in the lipid and fatty acid compositions of blood and other tissues
Linolenic acid (*n*-3)	Neurological symptoms: numbness; paresthesia; weakness; pain in limbs; blurring of vision

summary of these deficiency signs. In man, vitamin A deficiency disease is all too common in many parts of the globe (see Chapter 7) and vitamin D deficiency may occur in people in which a chronic lack of sunlight is combined with a poor diet, while overt essential fatty acid deficiency of a kind well documented in experimental rodents is probably relatively rare.

Can we see corresponding signs of nutrient overabundance as implied by the far right of Fig. 8.1? The most obvious example is when an overabundant energy intake, of which fat may be a major contributor, is manifested in some people in the condition of obesity. A major public health concern is the contribution of this condition to excess morbidity and mortality. Overabundance of vitamins A and D can be life-threatening as discussed in Chapter 7.

For every individual, there is a level of intake of each nutrient that will ensure that his body chemistry functions at its most efficient, an important prerequisite for good health. The nutritional requirements for any given nutrient may also depend on that individual's physiological state. Thus nutrients are generally required in greater abundance when growth is particularly rapid, during periods of substantial endocrine change, during pregnancy and lactation and after injury or similar stress. Requirements for a particular nutrient cannot be assessed in isolation from the diet as a whole, since they may be influenced by the nature and quantities of other components of the diet. This concept applies equally to fats as to other nutrients and a good illustration is in the way that the requirements for and metabolism of *essential* fatty acids can be

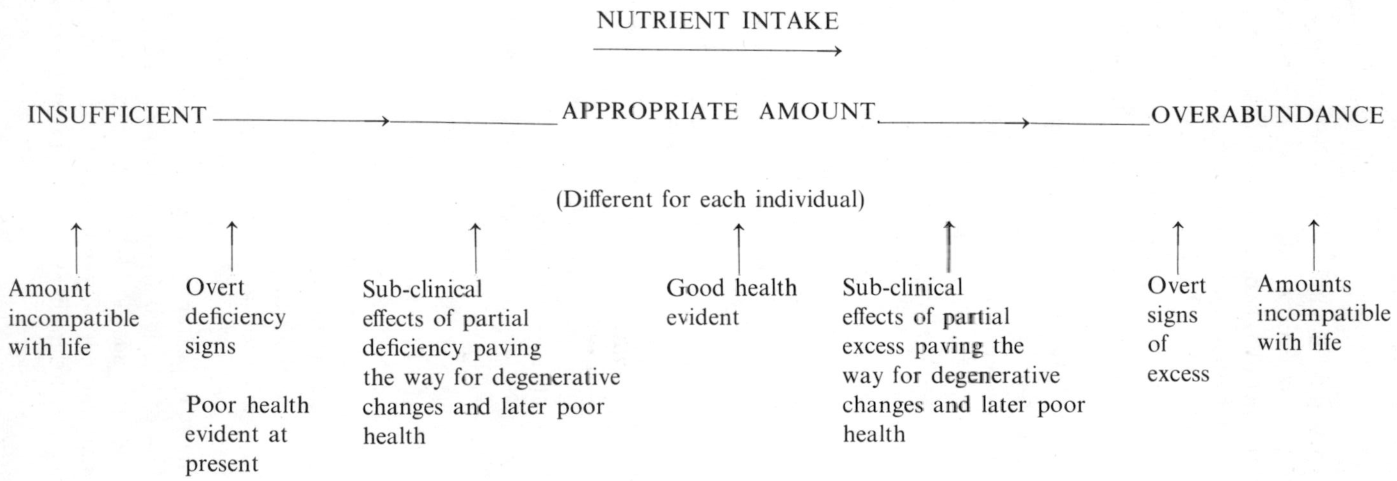

Fig. 8.1. Relationship between nutrient intake and body function ('health').

influenced by the types and amounts of *non-essential* fatty acids in the diet as described in Section 7.2.2. A huge intake of a particular type of non-essential fat, by competing for the enzymes involved in essential fatty acid metabolism, can alter the balance of important metabolic pathways with the possibility of adversely affecting health. Such subtle changes in metabolism almost certainly will not be immediately reflected in overt signs of disease. Rather will they result in slow degenerative changes, paving the way for later ill health over a time-scale measured in years rather than days. This situation is depicted in Fig. 8.1 on either side of the mid-point of 'appropriate' nutrient intake.

In contrast to this concept of 'balance' where subtle shifts in metabolism may be brought about by changes in proportions of nutrients, disease may be induced because particular dietary components (including certain fats) have specific toxic effects on tissues or because specific individuals have an 'inborn error of metabolism' which results in an inability to efficiently metabolize particular fats. In the latter case, particular fatty substances may accumulate in tissues, causing extensive damage. Section 8.2 will be devoted to brief descriptions of specific fats that have been claimed to have toxic effects and Section 8.3 will take the alternative approach of examining several diseases in which there is evidence for the involvement of fats as causal or aggravating factors. In these sections some attempt will be made to ask the question: Is there truly evidence for a direct toxic effect of the lipid or is this an example of a disturbance in nutrient balance? The answer to this question may be crucial in terms of being able to give appropriate dietary advice and in the formulation of national nutrition policies (see Section 8.4).

8.2. DIETARY FATS ALLEGED TO BE TOXIC OR TO HAVE TOXIC EFFECTS UNDER SOME CIRCUMSTANCES

8.2.1. Long Chain Monoenoic Fatty Acids

Some natural edible oils contain, in addition to the widely occurring monounsaturated fatty acid oleic acid (*cis*-9-octadecenoic), appreciable quantities of monounsaturated fatty acids with a chain length of 22 carbon atoms. The most important of these are rapeseed and mustard seed oils which contain erucic acid (*cis*-13-docosenoic acid) and herring oil which contains cetoleic acid (*cis*-11-docosenoic acid) (Fig. 8.2).

When marine oils containing polyenoic acids are used in margarine making, it is necessary to harden them by catalytic hydrogenation which

$CH_3\ (CH_2)_7\ CH = CH\ (CH_2)_{11}\ COOH$

(a) Erucic acid (*cis* - 13 - docosenoic acid)

$CH_3\ (CH_2)_9\ CH = CH\ (CH_2)_9\ COOH$

(b) Cetoleic acid (*cis* -11 - docosenoic acid)

Fig. 8.2. Long chain monoenoic acids.

gives rise to a range of long chain *cis*- and *trans*-monoenoic acid isomers. There has been some concern about the safety of fats containing these fatty acids and many feeding experiments have been performed with several animal species to assess their biological effects. For example, when young rats were fed diets containing more than 5% of the energy as rapeseed oil (45% erucic acid) their heart muscle became infiltrated with fat. After about a week, the hearts contained three to four times as much fat as normal hearts and although, with continued feeding, the size of the fat deposits gradually decreased, other pathological changes were noticeable, such as the formation of fibrous tissue in the heart muscle. The biochemistry of the heart tissue was also adversely affected in that the rate of biological energy-producing reactions was slower. The activity of lipases is lower with triacylglycerols containing erucic acid than with fatty acids of more normal chain length and this may reduce the rate at which these fats can be broken down. A review of the many experiments that have been carried out is quoted in the Bibliography. Although there are wide differences between species and the response is much weaker in older than in younger animals, the results have been taken very seriously from the point of view of human nutrition, as rapeseed oil is now a major edible oilseed crop on world markets.

It is not known whether similar lesions occur in man, although in some countries rapeseed oil has been consumed for many years, albeit not in the high concentrations employed in the animal feeding experiments. The toxicity is likely to be less in a mixed diet with adequate quantities of essential fatty acids and a good balance of the more usual fatty acids. Despite the lack of evidence for harmful effects in man, it has been thought prudent to replace older varieties of rape, having a high erucic acid content, with new varieties of zero-erucic rapes. Food products containing hardened fish oils will, however, continue to contain some C_{22} monoenoic acids and it might be wise not to let these (or any other single type of fatty acid) contribute too great a proportion of the diet.

8.2.2. Fatty Acids with *Trans* Unsaturation

Most unsaturated fatty acids found in nature contain double bonds in the *cis*-geometrical configuration. As explained in Section 3.3, fatty acids with the opposite (i.e. *trans*) geometrical configuration are found in nature, though in much less abundance (Figs 2. 2(g) and (j)). *Trans* fatty acids are found naturally either as short lived intermediates in biochemical pathways or as stable end products. As examples of the latter, *trans*-3-hexadecenoic acid is a ubiquitous constituent of photosynthetic tissue, though present in small amounts. Some seed oils (e.g. tung *Aleurites fordii*) may contain considerable amounts of *trans* fatty acids although they are not generally important dietary sources. *Trans* fatty acids are also produced by the process of microbial biohydrogenation of dietary polyunsaturated fatty acids in the rumen of ruminant animals (Section 3.2.1) or by chemical hydrogenation in margarine manufacture as described in Section 3.3.

In terms of man's diet, *trans* fatty acids may be consumed, therefore, in green vegetables (very small amounts), some seed oils, dairy products (up to 20% of the monounsaturated fatty acids in butter fat) or in margarine spreads and processed cooking fats. The *trans* fatty acid content of margarines may vary considerably ranging from 0 to 40–50% (Table 8.2).

Worries about the adverse nutritional effects of *trans* fatty acids began with the publication of results from an American laboratory which suggested that pigs fed hydrogenated vegetable fat for eight months had more extensive arterial disease than those on a control diet. More recent epidemiological studies have suggested that there is a correlation between the mortality from arterial disease and the consumption of hydrogenated fats in the UK.

Is there any evidence that the manner in which dietary *trans* fatty acids

TABLE 8.2
The Frequency of Distribution of *Trans* Fatty Acids in Margarines on Sale in North America

	% trans				
	0–10	*10–20*	*20–30*	*30–40*	*40–50*
% Frequency	12	35	27	22	4

Adapted from G. J. Brisson, *Lipids in Human Nutrition*, 1981, MTP Press, Lancaster.

are metabolized in the body could give rise to toxic effects arising from consumption of these fats?

There is no evidence that the efficiency of digestion, absorption and subsequent oxidation of fatty acids to yield metabolic energy is in any way disturbed by the presence of *trans* double bonds. When *trans* fatty acids are included in the diet, they can be found in the lipids of most body tissues. The most widely used experimental model has been the laboratory rat but some work has been concerned with the pig and man. The highest proportion of *trans* fatty acids found in human biopsy or necropsy specimens have been in liver and adipose tissue (up to 14%) but up to 0·5% of the fatty acids in brain may have the *trans* configuration. *Trans* monoenoic acids have also been found in human milk (0·1–4·5%).

In general, the incorporation of *trans* monoenoic fatty acids into tissues is proportional to the amounts present in the diet, but there is some evidence that the maximum level of *trans*, *trans*-linoleic acid is 6–9% regardless of the concentration of this acid in the diet or the duration of feeding. Moreover, the degree of incorporation is dependent on other constituents of the diet. The presence of even minimal levels of essential fatty acids in the diet tends to reduce the accumulation of *trans* fatty acids in tissue lipids.

There is some selectivity with regard to the fats into which *trans* fatty acids are incorporated. In general, there is a preferential incorporation of *trans* monoenoic acids into triacylglycerols because of the extensive deposition in adipose tissue. In other tissues, such as heart, liver or brain, *trans* fatty acids are more likely to be found in phospholipids, where they behave like saturated fatty acids in that they are preferentially located at position 1 in contrast to oleic acid which is more randomly distributed.

The metabolism of fatty acids in tissues is a dynamic process and once incorporated into tissue fats, they do not remain there permanently. Several weeks after changing from a diet rich in *trans* acids to one containing none, only negligible amounts are left in the tissues of experimental rats demonstrating the ready catabolism of *trans* fatty acids and their removal from tissues. *Trans* fatty acids may influence the metabolism of other unsaturated fatty acids. This is largely a result of competition between the different fatty acids for a single desaturase enzyme as discussed in Chapter 7. Linoleic acid (the primary dietary essential fatty acid) has the strongest affinity for this enzyme and is normally present in sufficient quantity to be used preferentially in the metabolic pathway, thus ensuring more than adequate levels of polyunsaturated fatty acids in the tissues for prostaglandin formation.

However, when the intake of linoleic acid is low, the presence of large quantities of fatty acids, such as *trans* monounsaturated fatty acids that can act as alternative substrates for the desaturase, influences the direction of the metabolic pathway so that the metabolic products are non-essential fatty acids that either cannot give rise to prostaglandins or result in the formation of prostaglandins of unknown or unpredictable activity. Thus the ingestion of excessive amounts of hydrogenated polyunsaturated fatty acids can result in metabolic and nutritional disturbances in two ways: the dietary intake of polyunsaturated fatty acids is reduced and the pathways for the metabolism of available essential fatty acids are disturbed by the unbalanced mixture of fatty acids ingested. Experiments with rats have shown that increasing the concentration of dietary *trans* fatty acids causes a decrease in the concentration of essential fatty acids in heart muscle and that this effect is accentuated when the diet is marginal in its essential fatty acid content.

It has also been suggested that an important factor in the toxicity of *trans* fatty acids is their effect in limiting the availability of the essential fatty acids for prostaglandin formation, either by displacing them from tissues or by inhibiting the metabolic pathways leading from essential fatty acids to prostaglandins. In one experiment, groups of rats were fed one of four diets containing 5% by weight of fat which was either (1) hydrogenated coconut oil (containing a negligible quantity of unsaturated fatty acids); (2) *all-trans*-linoleate; (3) an equal mixture of *all-cis*-linoleate and *all-trans*-linoleate; and (4) *all-cis*-linoleate. Diets containing *trans* fatty acids resulted in lower concentrations of essential fatty acids in the liver and in the blood platelets and a slower rate of production of prostaglandins by platelets. The differences were most marked when diets (1) and (2) were compared with (4); there were smaller differences between (3) and (4). The authors concluded that 'These data demonstrate that dietary *trans* fatty acids aggravate the symptoms of essential fatty acid deficiency and cause a reduction in the concentration of prostaglandins and their precursors, even when fed with *cis*-linoleic acid. The implications of these observations are significant in view of the presence of *trans* fatty acids in the American diet and are particularly pertinent because of the potent effects of prostaglandins and their intermediates on many physiological functions, particularly platelet aggregation and cardiovascular actions.' It is always difficult to translate results with animals under controlled experimental conditions into terms that are relevant to practical human nutrition and there are several reasons why we might wish to interpret these results cautiously. Diets (1) and (2) were not simply essential fatty acid deficient: they contained no essential fatty acids at all. It is well-established that the presence of *trans* or even *cis* monounsaturated fatty acids would exacerbate the effects of essential fatty acid deficiency under these conditions. Although diet (3) contained *all-cis*-linoleate, there was a slight depression in tissue essential fatty acid concentrations and a rather bigger depression of prostaglandin production. However, it is not established whether depressions of prostaglandin formation of this order are significant in the diet of man or indeed whether the high rate of prostaglandin production from diet (4), in which all the fatty acid was present as *cis,cis*-linoleate, can be regarded as normal. Finally, there is experimental evidence to show that the metabolic effects of *trans*, *trans*-linoleic acid are more potent than those of the *cis*, *trans* mixtures. Under modern processing conditions, however, the *trans*, *trans* isomer is a very minor component of diets and its metabolic effects may not be at all important in practice.

The problems that arise in interpreting the results of experimental work on *trans* fatty acids are illustrative of the difficulties of translating the results of animal nutrition studies to practical human nutrition generally. How representative is one species of another? What concentrations in the diet are appropriate in the test experiments? Scientifically, there may be merit in examining the effects of pure isomers included in a chemically well-defined diet. Practically, human beings eat a poorly defined mixture of substances in the presence of a very complex diet. What should the control diet be, against which comparisons are made? Some authors have used as a control fat, coconut oil, which is now recognized to have a particularly potent effect in raising blood cholesterol concentrations. Many experimentalists have employed feeding regimes in which concentrations of essential fatty acids were barely adequate. Because of the competition between dietary non-essential fatty acids and essential fatty acids (see Chapter 7) it is absolutely vital that feeding experiments are carried out with diets that have a sufficient inclusion of *cis*, *cis*-linoleic acid.

It is likely that the experiments that demonstrated increased arterial lesions in pigs fed *trans* fatty acids can be explained by a relative dietary deficiency of linoleic acid since another experiment from the same laboratory, in which diets were adequate in their content of linoleic acid, was not able to make a distinction between the pathological effects of *trans* and *cis* fatty acids.

The most reasonable conclusion from published data is that any adverse effects that might have appeared to result from the ingestion of fats containing *trans* fatty acids was not due to an effect of the *trans* double bond specifically, but to disturbance of essential fatty acid metabolism that occurs when the balance of non-essential to essential fatty acids is inappropriate. This may occur with a wide variety of non-essential fats (including *cis* monoenoic and saturated fats) especially when a single type of fatty acid dominates the diet. The golden nutritional rule of ensuring a varied diet, so that any one component does not outweigh any other is as applicable to the nutrition of fats as to any other nutrient.

The question now arises: Are diets that are commonly consumed likely to be unbalanced in this way? In Chapter 5, the changing patterns in consumption of food fats were discussed and in view of the trends towards greater consumption of those fats that are relatively rich in *trans* fatty acids (Table 8.3) it is inevitable that the consumption of *trans* fatty acids has been increasing. A similar trend has occurred in other European and North American countries and the Canadians have probably been foremost in documenting these changes. The annual

TABLE 8.3
Some Foods Containing *Trans* Fatty Acids

Food	*Total fat content (% by weight of food)*	trans *fatty acids (% of total fatty acid)*
Bread and rolls	2	10–28
Cakes	11–26	10–24
Instant and canned puddings	3–18	31–36
Fried potatoes (chips, french fries)	7–15	5–35
Potato crisps and similar snacks	36	14–34
Butter	82	1– 7
Hard margarines	81	18–36
Soft margarines	81	0–21
Shortenings	99	13–37

Adapted from G. J. Brisson, *Lipids in Human Nutrition*, 1981, MTP Press, Lancaster.

consumption of margarines and shortenings in Canada, for example, increased from 7 to nearly 15 kg per head between 1950 and 1980.

Calculation of the intake of *trans* fatty acids is extremely difficult, because of the variability in composition of different food items and because of the relative inaccuracy of food intake data for populations. Calculations based on a knowledge of the average *trans* fatty acid content of different foods and the trends in food consumption indicate average intakes of 12 and 9·6 g per head per day for the USA and Canada respectively, about 5% of which comes from animal fat and 95% from particularly hydrogenated vegetable oils. *Trans* fatty acids represent about 8% of the fatty acids in the diet.

A recent survey by the Ministry of Agriculture, Fisheries and Food has indicated that the intake of *trans* fatty acids in the average British diet is 8 g per head per day or $7\frac{1}{2}$% of the total fatty acid intake. A rather greater proportion (46%) of this comes from animal (mainly dairy) products in the UK than in North America (Table 8.4). As discussed in

TABLE 8.4
Estimated Average Daily Intakes of *Trans* Fatty Acids in the UK

Food	*Daily intake of fat from food (g)*	*Average* trans *fatty content of food (%)*	*Daily intake of* trans *fatty acids (g)*
Milk	16	6	1·0
Butter and other dairy products	28	6	1·6
Margarine and other fats and oils	33	9	2·8
Meats, fish, eggs	31	3	1·0
Cereal-based foods	12	11	1·3
	120		7·7

Adapted from R. Burt and D. H. Buss, Dietary fatty acids in the UK, *J. Clinical Practice*, 1984, in press.

The figures of Burt & Buss have been increased by a factor of 1·6 to take account of the following:

(a) The average daily fat intake is greater than their figure of 97·1 g because of food eaten outside the home. A figure of 120 g day^{-1} has been assumed.

(b) Most foods contain polyunsaturated fatty acids containing *trans* double bonds. Burt & Buss quote figures only for *mono* unsaturated acids.

Chapter 5, there is no evidence that the proportion of essential fatty acids in the British diet has been decreasing during this time; in fact, the reverse is true. Therefore, problems of toxicity of *trans* fatty acids should not arise. However, it should be emphasized that we are talking in terms of average intakes, and that some individuals will be taking in relatively more *trans* acids and fewer essential fatty acids than average and for these people there might be cause for concern. There might, therefore, be merit in clearly labelling the *trans* fatty acid content of fat products to aid in food selection. Such a system would also have the advantage of removing a current source of ambiguity: fats containing a high proportion of *trans* fatty acids can still be legitimately described as 'unsaturated' although in their physical and metabolic properties they much more resemble fully saturated fats.

8.2.3. Cholesterol

There is much confusion in the minds of many laymen (and scientists!) about the role of cholesterol. It is a word much bandied about, with little understanding of what it is. It has 'bad' associations and many people suppose that foods containing it should be avoided at all costs.

Cholesterol (Fig. 2.1(l)) is a prime example of a natural substance that

---

Some quantitative facts about cholesterol in the human body. The average adult human body contains rather more than 100 g of cholesterol which is present either as the free sterol or as cholesterol ester. It is found in all the membranes of all tissues and organs, in the fat stored in adipose tissue and in transit in the blood as lipoprotein cholesterol.

Some comes from the diet, some from synthesis within the body. Most tissues are able to synthesize cholesterol but the rates at which they carry out this synthesis vary considerably from tissue to tissue and also depend on the extent to which dietary cholesterol depresses tissue synthesis. There is continual exchange of cholesterol between the cholesterol in blood and the cholesterol in tissues. The rates of exchange are usually expressed in 'half-lives', that is, the time taken for half the substance in a given tissue to exchange with that in another 'pool'. The half-life of cholesterol tends to be short in tissues such as liver and intestine (a matter of hours), with slower rates of exchange in muscle and adipose tissue and even slower in the arterial wall.

According to some experimentalists, the body may synthesize about $\frac{1}{2}$ to 1 g of cholesterol per day, whereas a $\frac{1}{4}$ to $\frac{3}{4}$ g are taken in from the diet, of which only about 50% may be absorbed. Thus the average adult human body is making about 2 to 4 times as much cholesterol as is taken in from the diet in order to satisfy the body's demands for maintaining membrane structure and for providing bile acids and steroid hormones.

is an essential metabolite but not an essential nutrient. It is an essential metabolite because without it biological membranes would not function properly. No closely related steroid can, apparently, substitute entirely for cholesterol and still maintain essential functions. It is also important, as discussed in Chapter 6, in so far as it is converted into a variety of steroid hormones with important regulatory functions and into bile acids that are crucial to the absorption of fat. It is not an essential nutrient in so far as the body can manufacture its own supply if none is present in the diet (this is in contrast to linoleic acid). Among the reasons for its 'bad press' is that when plaques build up on the inner walls of the arteries in the disease known as atherosclerosis (onc of a number of degenerative diseases collectively known as cardiovascular disease — see Section 8.3.7), cholesterol and its esters are among the many substances contributing to the plaque. The cholesterol in the plaques has been shown to originate from cholesterol carried in the bloodstream in the lipoproteins mainly of the low density class. Within any population of people one can expect to find a very wide range of blood cholesterol concentrations. Exceptionally high concentrations are indicative of a metabolic disorder (sometimes inherited) that results in a failure to remove lipoproteins at a sufficiently rapid rate, an overproduction of lipoproteins by one of the tissues or a combination of the two (see also Sections 8.3.2. and 8.3.7). This is not at all the same thing as saying that high blood cholesterol concentrations are the primary cause of the disorder and this point has often been misunderstood by many people. Thus, although it is possible to influence blood cholesterol concentrations by dietary means, it does not follow that this will have any significant effect on the underlying disease if blood cholesterol concentration is only a secondary manifestation of a more deepseated disorder.

Contrary to widespread belief, changing the amount of cholesterol in the diet has only a minor influence on blood cholesterol concentration. Firstly, there is a limit to the amount of cholesterol that can be efficiently absorbed (Section 6.2) Secondly, the control mechanism described in Section 6.3 ensures that blood concentrations are maintained within certain limits by regulating the amounts taken into tissues and the rate at which cholesterol is synthesized by the body. Some individuals may, however, be more 'sensitive' to dietary cholesterol, because of the way in which these control mechanisms are 'tuned' just as some species of animals, such as rabbits, are particularly sensitive to dietary cholesterol and have been used as experimental models for arterial disease for that reason.

A more potent dietary influence on blood cholesterol concentration is the relative proportion of polyunsaturated to saturated fatty acids. Experimentally, it has been shown that saturated fatty acids are about twice as potent in raising the blood cholesterol concentration as polyunsaturated fatty acids are in lowering it. A ratio of *cis*-polyunsaturated to saturated fatty acids of about 0·5 (often referred to as the *P/S* ratio) stabilizes blood cholesterol whereas a higher ratio would lower it and a ratio of less than 0·5 would increase it by a calculable amount. The total amount of fat in the diet seems to be less important than the *P/S* ratio. For example, in one study of American adult men to compare the effects of isocaloric diets containing different amounts of fat and different mixtures of fats all giving a *P/S* ratio of 0·5, similar serum cholesterol and phospholipid concentrations were recorded in all the treatment groups even though the total fat contents of the diets varied from 5 to 30% of the total food energy.

To complicate matters further, it is now known that the way in which the fatty acids are distributed among the three positions in the triacylglycerol molecule as well as the nature of those fatty acids also have an influence on blood cholesterol concentration. Normally, a diet in which butter contributes a large proportion of the fat would result, for many individuals, in a rise in blood cholesterol concentration. The fatty acids in butter and in many other food fats are not randomly distributed among the three positions: specific fatty acids tend to be located at specific positions. Most of the short and medium chain acids in butter are at position 3. When the fatty acids are randomized (evenly distributed among the three positions) which can be done by interesterification (see Section 3.3.4) the fat no longer exerts its hypercholesterolaemic effect when fed to human beings. The reason for this remains obscure, but the observation does indicate one way of reducing the hypercholesterolaemic potential of fats if that were thought desirable.

This section has not been about disease. Its aim has been to put in perspective the observations on the role of cholesterol and some of the dietary factors that affect its metabolism. In terms of nutrition it does not merit all the attention it has been given over the years for a number of reasons. Firstly, many other factors in the human lifestyle, among them smoking, stress and exercise can influence cholesterol metabolism, often more dramatically than diet. Secondly, the concentration of any substance in the blood (which is a medium for transporting molecules from one part of the body to another) is a reflection of many processes some of which determine its entry into the blood and others that determine its removal therefrom. A change in its concentration over a period of time cannot therefore be more than an indication that the balance of one of these many processes has also changed. Finally, the total plasma concentration of cholesterol is the sum of the concentrations in the different lipoprotein classes (Section 6.3). Since each of these has a different origin and function, the relationship between blood cholesterol and disease processes can only be sensibly assessed in terms of the metabolism of specific lipoprotein classes and this theme will be further developed in Sections 8.3.2. and 8.3.7.

It is worth mentioning here that during the last few years, several researchers have expressed the view that arterial damage observed as a result of feeding cholesterol rich diets to animals should be attributed to oxidation products of cholesterol and not to the cholesterol itself. Cholesterol is normally a rather stable compound and would not be expected to oxidize rapidly on storage. Nevertheless, some oxygenated derivatives have been detected in commercial preparations of cholesterol. When these are purified and fed to animals, they more readily give rise to pathological lesions than purified cholesterol. There is also a possibility that oxygenated metabolites of cholesterol known to be formed in the body could also act as toxic agents. More research is required on this subject before its practical nutritional importance can be assessed.

8.2.4. Oxidized and Polymerized Fats

The storage of fats containing appreciable concentrations of polyunsaturated fatty acids in the presence of oxygen at room temperature can result in the formation of hydroperoxides (see Sections 3.3.2. and 3.3.7.). When these are ingested they are rapidly degraded in the mucosal cells of the gut to various oxyacids that are further oxidized to carbon dioxide. There is no evidence for the absorption of unchanged hydroperoxides nor for their incorporation into tissue lipids. In one experiment, for example, the growth of rats fed fat with a peroxide value of 100 was reported to be normal. However, other workers have provided evidence for the potentiation by linoleic acid hydroperoxide of tumour growth in female rats (see also Section 8.3.6.). It should be mentioned that once the peroxide value exceeds 2–5, the fats are in any case organoleptically unacceptable to human beings.

It seems that the normal levels of autoxidation of foodstuffs at room temperature may not be important from a toxicological point of view. Heating unsaturated fats leads to more extensive chemical changes. Heated fats do not contain peroxides, but instead a range of polymerized compounds is obtained. The extent of this polymerization may be 10–20% under normal household or commercial practice, up to 50% when the oil is severely abused. Only in the latter case have toxic effects been observed in animal feeding experiments and then of no great severity. Some reviewers have concluded that fats heated in normal cooking processes are not harmful to man and subsequent studies have tended to support this contention (see Alexander in the Bibliography) although others take a less optimistic view (see Artman in the Bibliography). The potential toxicity of oxidized cholesterol has already been referred to in Section 8.2.3.

8.2.5. Cyclopropene Fatty Acids

By far the most important edible oil containing these fatty acids (of which sterculic acid is an example (Fig. 8.3)) is cottonseed oil in which the concentration ranges from 0·6 to 1·2%, although after processing, the oil as actually eaten probably contains only 0·1–0·5%. Animal fats contain no cyclopropene fatty acids. The metabolic importance of sterculic acid is to inhibit the desaturation of stearic to oleic acid, the effects of which are to alter the permeability of membranes or to increase the melting point of fats. The biological results of this are illustrated by the high melting point of lard in pigs fed cottonseed meal and 'pink-white' disease in hen's eggs. If cyclopropene acids are present in the diet of laying hens, the permeability of the yolksac membrane is increased, allowing the release of substances, including pigments, into the white. In dietary experiments with animals, the source of cyclopropenes is usually *Sterculia foetida* seed oil, which has a much higher content (up to 70%) than cottonseed oil. Rats die within a few weeks when fed diets containing 5% of dietary energy as sterculic acid and the reproductive performance of females is completely inhibited with levels as low as 2%. Man has been eating cottonseed oil for many years in such products as margarines, cooking oils and salad dressings but the intake of cyclopropene fatty acids (which in any case are unstable compounds) is very small. On this basis, it is presumed that low levels have no adverse effects, but whether prolonged ingestion by man of larger amounts would

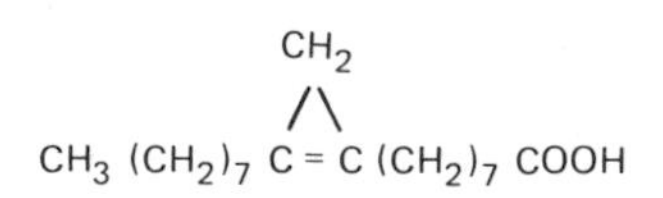

Fig. 8.3. Sterculic acid.

be deleterious is not known. The possibility of synergistic effects in which fats containing cyclopropenes may enhance the potency of carcinogens deserves more research attention.

8.3. DISEASES IN WHICH DIETARY FATS MAY PLAY A DIRECT OR INDIRECT ROLE

8.3.1. Lipid Storage Diseases

The assimilation of all nutrients requires them to be metabolized in some way. Most biochemical reactions would proceed very slowly indeed if it were not for the presence in all tissues of enzymes, i.e. proteins that increase the rates of biochemical reactions. Sometimes, the genetic blueprint that determines the biosynthesis of body proteins is faulty and a specific enzyme may be missing completely or working inefficiently. When this occurs, it may result in the inability to assimilate a particular nutrient or, once assimilated, to convert it into an essential body metabolite. Such genetic disorders or 'inborn errors of metabolism' are comparatively rare, but may pose a life or death nutritional problem for the sufferer.

Several inborn errors of metabolism exist in which the missing enzyme is one which is involved in the breakdown of a specific lipid molecule. Since the biosynthesis of these lipids is not impaired, the result of the enzyme deficiency is the gradual accumulation of the lipid in tissues. Most of the important diseases of this type are ones that involve structural lipids of the central nervous system and they are summarized in Table 8.5. The diseases are rare and frequently fatal, which serves to indicate how important it is that the amounts and types of lipids in membranes are strictly controlled in order to preserve biological function. Many of the lipids involved in these disorders are readily synthesized in the body, so that dietary treatment is not effective. However, there is one lipid storage disease, known as Refsum's disease, which can be controlled by strict exclusion of a specific fatty acid from the diet. This disease is the result of the failure to break down the branched chain fatty acid phytanic acid (Fig. 8.4). This fatty acid is formed from phytol, a universal constituent of green plants. As a result of the enzyme defect, phytanic acid accumulates in tissues, causing multiple defects such as night blindness, narrowing of the visual field, skeletal malformation and cardiac complications. The nervous system is affected since when the myelin membrane (which protects the nerve fibres) has

TABLE 8.5
Lipid Storage Diseases

Lipid accumulating in tissue	*Name of disease*	*Defective enzyme*	*Effects of disease*
Sphingomyelin	Niemann–Pick disease (sphingomyelinosis)	Sphingomyelinase	Deposition of sphingomyelin in almost every tissue; loss of function frequently fatal before third year
Gangliosides	Tay–Sachs disease	Terminal *N*-acetyl galactosamine cleavage enzyme	Accumulation of an abnormal ganglioside in tissue especially brain, impairment of mental and somatic function and vision; eventually demyelination
Glucosyl ceramide	Gaucher's disease	Glucocerebrosidase	Appearance of large lipid-laden cells in spleen, liver and bone marrow; pigmentation of skin
Trihexosyl ceramide	Fabry's disease	Ceramide trihexosidase	Skin rash; pains in the extremities, pyrexia; progressive renal failure
Sulphatide	Metachromatic leucodystrophy	Sulphatase	Impairment of motor function; ataxia; coarse tremor and progressive demyelination
Phytanic acid	Refsum's disease	α-Oxidation enzymes	Chronic polyneuropathy; night-blindness; narrowing of visual field; skeletal malformation

Reproduced from M. I. Gurr and A. T. James, *Lipid Biochemistry: An Introduction*, 1980, Chapman and Hall, London.

$$
\begin{array}{ccccccc}
 & CH_3 & & CH_3 & & CH_3 & & CH_3 \\
 & | & & | & & | & & | \\
CH_3\,.\, & CH(CH_2)_3\,.\, & & CH(CH_2)_3\,.\, & & CH(CH_2)_3\,.\, & & CH\,.\,CH_2\,.\,COOH
\end{array}
$$

Fig. 8.4. Phytanic acid (3, 7, 11, 15-tetramethylhexadecanoic acid).

accumulated a certain proportion of phytanic acid it becomes unstable and starts to disintegrate. The disease is normally fatal and to survive, patients must have a phytanic acid-free diet.

Branched chain fatty acids occur in ruminant fats (Fig. 2.2). An important precursor of branched chain fatty acids is methylmalonyl-CoA. This metabolite does not normally accumulate in tissues because it is removed by an enzyme that depends on a supply of vitamin B_{12} for its activity. A rare metabolic abnormality has been described in children in whom a congenital deficiency of B_{12} metabolism results in a failure to remove methylmalonyl-CoA. This in turn leads to the formation of branched chain fatty acids and their accumulation in tissues, causing some neurological disturbances. However, no overt neuropathological signs have been observed in young sheep and goats that were fed on barley-rich diets for up to 2 years despite their having synthesized considerable amounts of branched chain fatty acids in that period. Equally, no neurological disorders were evident in baboons which were deprived of vitamin B_{12}, even after the administration of an inactive analogue which exacerbated the depletion of B_{12} and which occasioned the hepatic production of enhanced amounts of branched chain fatty acids. Hence, it seems that the only reason for concern about the toxicity of lipids containing branched chain fatty acids is in the case of those few individuals with specific metabolic disorders that prevent the normal utilization of these compounds. For most people, the branched chain acids pose no problem. Indeed, there is some evidence that a branched chain fatty acid (14-methylhexadecanoic acid) may play an essential role in protein synthesis. These results need to be confirmed by other laboratories and future developments may indicate a vital importance for specific branched chain fatty acids, occurring in dairy products, in nutrition and metabolism.

Finally, several exceedingly rare triacylglycerol storage diseases have been described, resulting from a deficiency of the lipase in adipose tissue. It might be thought that obesity could be described as a triacylglycerol storage disease. There is, however, no evidence to suggest that the condition of human obesity that is widespread in affluent countries is due

TABLE 8.6
Diseases Involving Abnormal Serum Lipoprotein Concentrations — the Lipoproteinaemias

Classification	*Characteristic lipoprotein pattern*	*Major lipids involved*	*Treatment*	*Clinical features*
Hyperlipoproteinaemias				
Type I	Raised chylomicrons in fasting plasma	Triacylglycerols Free cholesterol	Restricted fat diet; replace long chain fats by medium chain triacylglycerols (MCT,)	Rather rare; usually diagnosed before age 10
Type II	Raised LDL but the LDL are normal in composition	Cholesterol esters	Restrict dietary cholesterol by limiting the dietary intake of egg yolks, liver, dairy products Substitute polyunsaturated oils, PUFA margarines, skimmed milk	A common disorder; very strong associations with premature heart disease (IHD); children who are homozygotes for the disorder exhibit features of IHD in the first decade of life; xanthomas (massive accumulation of cholesterol in the skin) and corneal arcus (white ring in the eyes) are common features of the disease; occurs both as a heritable disorder and secondary to hypothyroidism
Type III	Raised abnormal LDL concentrations	Cholesterol esters Triacylglycerols	(i) Restrict cholesterol intake (ii) Reduce weight (iii) Dietary composition should be protein:carbohydrate:fat = 20:40:40	The third most common disorder after types II and IV; accompanied by extensive vascular disease

			(iv) Drug treatment with clofibrate, 1·5–2 g day^{-1}	
Type IV	Elevated VLDL concentrations	Triacylglycerols	(i) Weight control (ii) Avoidance of excessive dietary carbohydrate (iii) Hypolipidaemic drugs (e.g. Clofibrate)	Associated with abnormal glucose tolerance and a family history of diabetes; obesity is extremely common; not associated with IHD to the same extent as type II; occurs both as a heritable disorder and secondary to diabetes pancreatitis, etc.
Type V	Elevated chylomicrons and VLDL	Triacylglycerols Cholesterol esters	(i) Weight reduction (ii) Low energy diet not rich in either carbohydrate or fat: difficult to achieve	Very rare
Hypolipoproteinaemias (*very rare*)				
Familial LDL deficiency	Deficiency or complete absence of LDL; poor ability to form chylomicrons after a fatty meal	Cholesterol esters Triacylglycerols	Limit long chain saturated fat intake and replace by MCT with some PUFA vegetable oils	Neuromuscular disturbances; retinal changes; red blood cell abnormalities; steatorrhea (bulky and excessively fatty stools)
Familial HDL deficiency (Tangier disease)	Abnormally low HDL concentrations	Cholesterol Phospholipids	—	Enlarged tonsils, spleen, liver and lymph nodes; accumulation of lipids in reticulo-endothelial tissues

Reproduced from M. I. Gurr and A. T. James, *Lipid Biochemistry: An Introduction*, 1980, Chapman and Hall, London.

to a specific enzyme defect. It results from a failure adequately to balance energy expenditure with energy intake and will be allotted to a separate section (Section 8.3.3).

8.3.2. Hyper- and Hypo-lipoproteinaemias

Just as enzyme defects can result in diseases characterized by the inappropriate accumulation of fats that normally fulfil a structural role, as described in the last section, so there may be metabolic defects affecting the metabolism of transport lipids. The result is abnormally high concentrations (hyperlipoproteinaemia) or low concentrations (hypolipoproteinaemia) of lipoproteins in the blood plasma. The disorders are usually classified as 'primary' or 'secondary'.

The *primary* disorders are normally inherited and may be due to the lack of an enzyme that removes lipoproteins from the plasma, a defect in the cell membrane receptors that control their uptake from plasma into tissues (see Section 6.3) or a defect in a key pathway of lipoprotein metabolism, possibly in the liver. The major lipoproteinaemias are summarized in Table 8.6. Nutrition is important because in the case of an enzyme defect, certain fats may have to be eliminated from the diet or at least severely restricted. Thus, type I hyperlipoproteinaemia resulting from a defect in lipoprotein lipase, can be controlled by restricting the intake of triacylclycerols containing long chain fatty acids and substituting medium chain triacylglygerols. In type II hyperlipoproteinaemia, which is characterized by abnormally high concentrations of low density lipoproteins, the intake of dietary cholesterol and of saturated fats has to be restricted and the *P/S* ratio raised by the inclusion of more polyunsaturated fats in the diet. Extensive vascular disease occurs within the first few years of life and to increase life expectancy, drug therapy often has to be combined with dietary management. Among the drugs used are resins that very strongly bind to bile salts in the gut, thus increasing their excretion in the faeces. This has the effect of increasing the elimination of bile salts and therefore cholesterol through the entero-hepatic circulation.

The *secondary* hyperlipoproteinaemias occur when the metabolism of lipoproteins is altered as the result of another recognizable disease which, if treated, will lead to a normalization of the lipoprotein pattern. Examples are diabetes and obesity which are frequently accompanied by abnormal lipoprotein patterns characteristic of type IV hyperlipoproteinaemia (Table 8.6). Normalization of glucose tolerance and of weight, respectively, results in a normalization of the lipoprotein pattern and this theme is developed in the next two sections.

8.3.3. Obesity

Obesity is one of the predominant health problems in affluent societies. It has been the subject of much research but little progress has been made towards its control. This is probably because, although there is undoubtedly a metabolic component that is amenable to research, there are also important psychological and sociological components that are different for each obese individual.

The most obvious implication of fats in obesity is their accumulation in adipose tissue to an inappropriate degree. This is, however, only the outward sign—the final stage in a series of metabolic changes leading to obesity. The cause is not necessarily to be found in adipose tissue metabolism.

An individual is said to be in energy balance when his fat stores are neither expanding nor contracting; in other words, energy intake in the form of food is equally balanced by energy expenditure, as discussed in Section 5.2.2.

Because dietary fats contain over twice as much energy per gram as carbohydrates and proteins, high fat diets, common in many Western countries have often been assumed to play a role in the development of obesity. One may find populations, however, in which high fat diets are widely enjoyed, yet which do not have a high prevalence of obesity and some individuals may be fed experimental diets containing more than enough fat to supply their energy requirements several times over, yet not gain weight (Section 5.2.2). We should not confuse the possible role of fat in the aetiology of the disease, with its place in the management of obesity once acquired. Whereas fat may have played no great part in the processes leading to the obese state, it may well be prudent for those seeking to lose or maintain weight, to avoid too much dietary fat on account of its high energy density.

There is a potential for food technology to come to the aid of people who are struggling to control their energy intake. The concept of a low fat spread is that for the same bulk of the food, the energy content can be reduced by about a half. These low fat spreads will only be effective in human diets if there is no tendency (as there is in the rodent) to compensate for the low energy density of the food by automatically adjusting total food intake over the medium or even the long term. Surprisingly little work has been done to test this hypothesis with human subjects and it deserves further attention. 'Tailor-made' fats that mimic the physical and organoleptic properties of natural fats but which resist digestion and therefore have little or no energy value, may have a role in energy reduced diets. Examples are the glycerol polyesters. They have not yet found widespread use either for reasons of cost or because the presence of large quantities of undigested fat in the gut causes physiological problems. Few rigorous studies on the energy value of these synthetic fats have been done and before they could be used extensively in foods, their safety in use would need to be extensively assessed. Their potential for incorporation into foods to increase the potential range and variety of low energy diets deserves further research. The potential for manipulating the energy expenditure side of the energy equation should not be forgotten. Although there is little evidence to suggest that the efficiency of use of polyunsaturated fats is significantly different from that of saturated fats (Section 5.2.2), there is no doubt that the medium chain fats are metabolized by oxidation rather than being stored in adipose tissue (Section 5.2.2) and these fats could find wider dietetic applications.

8.3.4. Diabetes

Diabetes is another common disease of affluent countries the occurrence of which is slowly but gradually increasing. The term is a cause of confusion since it is applied to at least two diseases that are quite different in their associated metabolism. In 'juvenile onset' diabetes, the cells of the pancreas that normally produce insulin are defective and the patient is unable to maintain normal control of blood glucose concentration. The condition is often associated with leanness and a reduced level of fatty acid synthesis. It is controlled by administration of appropriate amounts of insulin and control of the diet to maintain stable blood glucose concentrations.

'Maturity onset' diabetes is invariably associated with obesity, hyperlipoproteinaemia and an increased tendency to develop atherosclerosis. In contrast to the juvenile type, fat synthesis seems to be enhanced and blood glucose *and* insulin concentrations are elevated. The condition is associated with a decreased sensitivity of tissues, such as adipose tissue, to insulin; in other words, more hormone than normal is required to produce a given metabolic effect. The pancreas responds by producing more insulin which in turn enhances the general level of lipid synthesis. Since insulin is not lacking, the management of the disease is predominantly by means of dietary modification.

Whereas diabetes was once regarded as a disease of carbohydrate metabolism, it is now apparent that faulty fat metabolism is equally important. The main practical implication is a radical change in the dietary management of diabetes. Extremely low carbohydrate diets are no longer emphasized to the extent they used to be although it is recommended that any increase in carbohydrate should contain a higher proportion of complex carbohydrate (starch) and of non-digestible polysaccharides (dietary fibre). The chief measures are to reduce total energy intake to avoid obesity and to increase the proportion of polyunsaturated to saturated fat to control hyperlipoproteinaemias.

8.3.5. Fats and Immunity

Strictly speaking, this section is not concerned with a specific disease entity. However, as the course of many diseases is determined by the strength of the body's immune defences and there is evidence to link some facets of immunity with dietary fat, it is appropriate to devote a section to this subject.

In the well-nourished and healthy animal the body is able to resist attack by infective agents by mobilizing the resources of the immune system. This may involve the production of specific antibodies that recognize infective organisms or other 'foreign' materials (antigens) or the activation of a protective network of special cells called 'lymphocytes'. These different aspects of the immune system are termed 'humoral' and 'cell-mediated' immunity, respectively.

In man, protein energy malnutrition markedly depresses the cell-mediated immune system increasing the risk of infection, whereas the humoral immune system is less severely impaired. In addition to the well-documented effects of protein energy malnutrition, deficiencies of zinc, vitamin A and essential fatty acids are also associated with malfunction of the immune system.

Just as malnutrition is associated with an impaired ability to mount an adequate immune response, overnutrition also seems to affect immunity adversely. Most of the evidence for this comes from observations of experimental animals and is somewhat indirect. Some strains of mice, for example, have an inherited tendency to autoimmune disease; that is, their immune defence system reacts against one of the body's own proteins, mistaking it as foreign. When the energy consumption of these animals is restricted, there is an improvement in virtually all aspects of immune function and the animal's lifespan is significantly increased. Other diseases in which impaired immunity is thought to play some role, such as cancer, are less severe when diet is restricted and obesity avoided. There is some evidence that the source of dietary energy most important in exerting this suppression of immunity is the fat component. Observations that support this contention are first that diets supplemented with polyunsaturated fats such as sunflower seed oil, seem to aid treatments designed to reduce the tendency of the immune system to reject organ transplants. They have also been found to be beneficial, in some dietary studies, in reducing the severity and frequency of relapses suffered by patients with multiple sclerosis, a disease in which autoimmunity is thought to play a part.

Experimental animals of various species, when fed high fat diets (about 40% of energy) are less efficient at producing antibodies when immunized with antigens than animals fed diets containing only 4% of the energy as fat. They also show a weaker inflammatory reaction when the antigens are injected under the skin and the ability of their lymphocytes to respond to immunological stimuli is damped down by a component of their blood serum. This substance is a type of lipoprotein and current ideas are tending towards the belief that lipoproteins may play an important role in regulating the immune system. It is not yet clear whether the essential fatty acids have a specific role in this immunoregulation, or whether it is a property of fatty acids in general. We know that prostaglandins on their own can either suppress or enhance immune activity depending on the type of prostaglandin used in the test or its concentration. What is probably happening in the living animal is that the 'balance' of the immune system is responsive to the competing actions of several prostaglandins, whose production is regulated by the balance of essential to non-essential fatty acids in the tissues. This in turn is at least in part responsive to diet. Another effect of changing dietary fat may be to alter the composition of fatty acids in cell membranes, which could affect the physical properties of the membrane. By looking at Fig. 2.3 it can be appreciated that a change in the physical properties of the lipid bilayer could influence the disposition of proteins located in the membrane. Their shape, freedom to move and exact position in the membrane could be altered and this might affect their properties. Immune reactions depend on the antigenic protein interacting with a protein on the cell membrane called a 'receptor' and changes in shape during this interaction trigger off a series of chemical reactions that eventually leads to an immune response. The same principles apply to responses to hormones and to reactions catalysed by enzymes. The whole field of membrane chemistry and its response to nutrition are exciting new fields in which substantial advances can be expected in the next few years.

8.3.6. Cancer

Knowledge about the involvement of any particular nutrient in the development of a disease often comes from epidemiological studies. The epidemiologist collects statistics about the occurrence of a specific disease within a population and tries to relate them to certain characteristics of the lifestyles of people in that population. These techniques have brought to light some extremely interesting relationships between the consumption of certain foods and the prevalence of cancer in many parts of the world. A strong positive correlation between the daily intake of fats and oils and the age-adjusted mortality from cancer of the colon (large bowel) is shown by data collected from 21 different countries. Of particular interest is the low incidence of colon cancer in Japanese people living in Japan, where the amount of fat in the diet is small, compared with that of Japanese migrants to the USA whose daily intake of fat is much higher. Even in Japan, the incidence is gradually increasing as the diet becomes modified by Western influence and its fat content rises.

Cancer is an all-embracing term for many diseases each with a different demographic distribution which may give us clues about causes. Hence, colon cancer is prevalent in the fat-loving UK whereas liver cancer is not. The latter disease is prevalent in some parts of Africa where it can be associated specifically with the presence of aflatoxin, a highly carcinogenic substance produced by a mould that infests certain foods such as peanuts. Breast cancer incidence is strongly correlated with the total fat intake of populations, less strongly with animal fat intake and hardly at all with vegetable fat intake. The reasons for these correlations are discussed later.

The difficulties involved in drawing conclusions from epidemiological data are nicely illustrated by the recent debate about the inverse relationship between serum cholesterol concentration and the morbidity and mortality due to cancer. Analysis of several epidemiological studies throughout the world leads to the conclusion that cancer incidence is highest where serum cholesterol concentrations are lowest. It must be emphasized that a strong correlation of *A* versus *B* arising from an epidemiological study is in no way indicative that *A causes B*. In the specific case of serum cholesterol and cancer, it could well be that changes in serum cholesterol are a result of another biological event which itself is involved in causing the cancer. Most of the studies from which the data were collected were actually designed to collect data on heart disease and therefore the numbers of cancer cases were quite small especially when broken down into specific types of cancer. The results could be a result of competing lethal risks: if part of the population sample has already succumbed to ischaemic heart disease there may be a bias towards increased risk of other diseases in the remainder. By far the biggest problem in the epidemiology of cancer, or of any other slowly developing disease, is that one does not know whether the correlation existed at the time when the disease was initiated. The change in serum cholesterol might be the end result of a series of metabolic changes brought about by the cancer itself. If this could be proved it might act as a marker for the presence of a cancer before it had been chemically diagnosed. Alternatively, there might be a direct cause and effect relationship between the high serum cholesterol concentration and the development of the cancer.

While epidemiology cannot give direct evidence of cause and effect relationships and should not be used in this way, it is useful in giving the nutritionist, physiologist or biochemist clues about the most worthwhile avenues for research. Is there any experimental evidence to link any type of dietary or body fat with the development of cancer? The formation of tumours is thought to proceed through three distinct stages. The first stage is called *initiation* in which a chemical substance known as a 'carcinogen' (which may either originate from the diet or be produced during body metabolism) causes a permanent alteration to the genetic material in a cell. Typical initiator substances are frequently fat-soluble and the polycyclic hydrocarbon dimethylbenzanthracene (DMBA) is commonly used as an experimental carcinogen. The second stage is promotion in which another substance, the promotor, increases the chance of the misinformation, inserted into the genetic code by the initiator, being expressed. From experiments with laboratory animals, it has been deduced that polyunsaturated fatty acids can act as promotors by enhancing the development of mammary tumours initiated by DMBA. There is no evidence that they can act as initiators. It is not yet known what mechanism is involved in the promoting action of polyunsaturated fatty acids. They may exert their effect by suppressing the immune system (see Section 8.3.5) which tries to defend the body against the tumour cells. The third stage in tumour formation is called *propagation* which is any process that stimulates cells to grow. Polyunsaturated fatty acids could act as propagators by modifying the cell surface or taking part in and stimulating the process of cell division.

Some experiments in which the carcinogen DMBA was administered to rats fed diets containing different amounts and types of fatty acids throw some light on epidemiological associations between dietary fat and cancer incidence discussed earlier. It appears that there must be a minimal level of polyunsaturated fat in the diet for promotion to occur, but this will only be effective when the total fat content of the diet is high—about 40% of dietary energy. A low fat diet rich in polyunsaturated fat was ineffective in promoting the effect of DMBA as was a high fat diet that contained no polyunsaturated fatty acids. As most human diets contain at least enough essential fatty acids to protect against essential fatty acid deficiency, the main correlation observed is therefore between total fat intake and cancer incidence.

The experiments just described were concerned mainly with mammary cancer. One of the primary hypotheses for the development of colon cancer is that dietary fat influences the composition of the microflora of the lower bowel, encouraging the establishment of certain bacteria that convert bile acids to carcinogens. Epidemiologically there is a strong association between the consumption of animal fat and the faecal excretion of certain types of bile acids. Experimental rats fed on diets rich in animal fat were more susceptible to colon tumours induced by dimethylhydrazine and excreted greater quantities of certain steroids and bile acids than those fed a low fat diet. In animals reared in such a way as to ensure that the alimentary tract was free of microorganisms, the incidence of tumours was particularly low, providing evidence for a role of the gut microorganisms. Although it was stated that polyunsaturated fats do not act as tumour initiators, it could be that fats abused by heating so that they contain a high level of peroxides could themselves act as carcinogens and there is some evidence from animal experiments to support this. The occasional consumption of oxidized fat is likely to do little harm but the constant re-use of heated fats would be better avoided.

Epidemiological studies have indicated a positive association between low intakes of vitamin A or its precursors and increased susceptibility to cancer. Some experimental data with rats deprived of vitamin A lend support to this. In affluent countries, however, few people are deficient in vitamin A and in terms of therapy, interest is likely to centre on the pharmaceutical use of synthetic retinoids rather than natural vitamin A which in large doses is extremely toxic. In developing countries, the principal result of vitamin A deficiency is xerophthalmia which will continue to be a more severe problem than cancer.

8.3.7. Vascular Diseases

Diseases of the vascular system, cardiovascular and cerebrovascular diseases, currently account for a large proportion of the morbidity and mortality in industrialized countries. Much has been written and spoken about the role which nutrition might play as a factor in causing vascular disease or at least in influencing its course and there can be few people who do not have some impression that dietary fats are closely associated with heart disease for good or ill and that cholesterol might play a villainous if shadowy role in the drama! It would be impossible to give an adequately detailed and critical assessment of all the evidence for relationship between fats and vascular disease in a single section of a book of this type. What follows will be a brief and rather personal account of the various ways in which fats in the diet may influence vascular disease and an attempt to put some perspective into what many may feel is a scene of utter confusion.

The major cause of death is the disease known as 'ischaemic heart disease' (IHD) and the following discussion will be devoted almost entirely to IHD. It is worth mentioning, however, that vascular disease affecting other parts of the body, while not necessarily proving lethal, can be the cause of a great deal of illness and distress and should not be forgotten in the quite natural emphasis of IHD.

IHD is a disease in which the coronary arteries become narrowed by the accumulation of deposits to such a degree as to prevent the coronary circulation meeting the metabolic demands of the heart. The immediate cause of death is the failure to supply a sufficient proportion of the heart muscle with oxygen. In considering the role of fats, it is useful to consider the disease as consisting of two distinct phases. The first phase is known as *atherosclerosis*, an irregular thickening of the inner wall of the artery that reduces the size of the arterial lumen. The thickening is caused by the accumulation of 'plaque', consisting of smooth muscle cells, connective tissue and considerable deposits of fats of which cholesterol esters comprise a major part.

The second phase, and the fatal episode in IHD, is the formation of a thrombus or blood clot in a coronary artery, usually at the site of an atherosclerotic lesion so that the blood supply to the heart is cut off (*myocardial infarction*).

Most working hypotheses of atherosclerosis, which is present in most human beings to some degree, even from an early age, although clinical signs may not be apparent, assume that the initiating step is damage to the vessel wall. It is not fully understood what causes the damage in practice, although arterial lesions can be initiated experimentally by a number of chemical agents including cholesterol or oxidation products of cholesterol. It has also been suggested that an immunological reaction to a food protein or an autoimmune reaction might be the initiating event and that high blood pressure is a necessary concomitant for damage to occur. The site of damage becomes a focus for the proliferation of smooth muscle cells and macrophages and this mass of cells has been likened to a tumour by some researchers. The cells accumulate large amounts of fats, and by labelling the fats carried by serum lipoproteins with radioactivity, it can be shown experimentally that at least some of the fat in the atherosclerotic plaques can originate from the serum lipoproteins. This is one of several reasons why a link has been made between the concentrations of fats (or, strictly speaking, lipoproteins) in the blood and the development of fatty plaques.

The sequence of associations

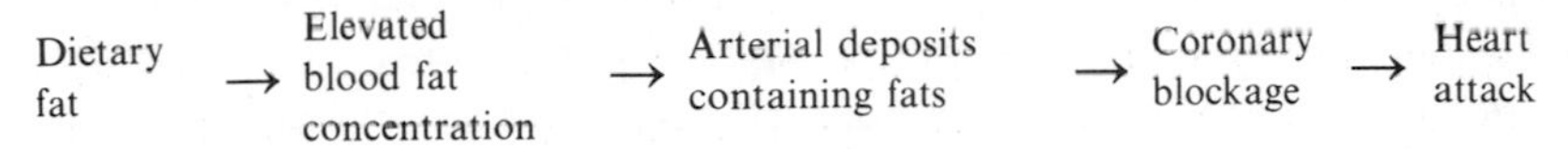

or variations of it, have come to be known as 'the lipid hypothesis of heart disease' and a rather more detailed scheme of associations and interactions is illustrated in Fig. 8.5. Experimental and epidemiological evidence exists for individual parts of the sequence but this is not the same thing as saying that the entire chain of events is the major contributor to the human disease.

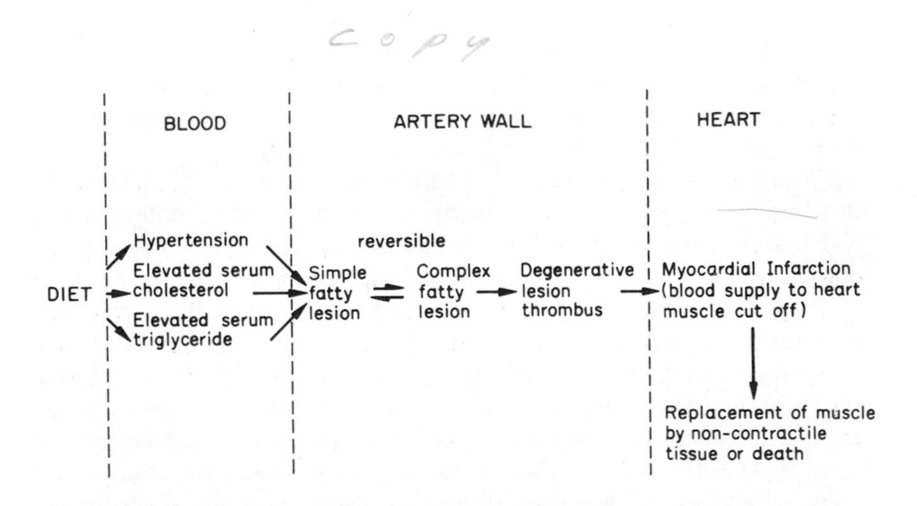

Fig. 8.5. Possible processes involved in the development of ischaemic heart disease. (Reproduced from M. I. Gurr and A. T. James, *Lipid Biochemistry: An Introduction*, 3rd Edn, 1980, Chapman and Hall, London, by kind permission of the publishers.)

First of all, there is epidemiological evidence from many studies around the world to relate the intake of dietary fats, particularly those in which the *P/S* ratio is rather low, with high serum cholesterol concentrations. Similarly, these studies have generally, though not always, demonstrated a relationship between the average serum cholesterol in a population and the risk of developing heart disease. At our current state of knowledge, it is important to talk in terms of low density lipoprotein cholesterol since it is now known that high density lipoprotein cholesterol is inversely related to the risk of heart disease. Many quoted studies were done before the importance of lipoproteins was realized. As in cancer epidemiology, powerful evidence about dietary relationships comes from studies of migrant populations. Thus, there has been an increase in the incidence of heart disease in Japanese who emigrated to the USA from their own country where the incidence is low and the fat intake is low with a high *P/S* ratio. Their blood cholesterol pattern changed at the same time from a characteristically low Japanese level to one more typical of the USA. The association between blood lipid concentration and the incidence of IHD is much weaker in individuals within a population implying that whereas, on average, high serum low density lipoprotein (LDL) concentrations indicate increased risk of the disease, it may not be true for a particular individual. Other factors of environment and lifestyle may exert a greater influence. Important ones are said to be smoking, exercise and hypertension and it is important to realize that these factors themselves can be independently associated with high blood lipoprotein concentrations.

Those who argue against the lipid hypothesis particularly against the idea that dietary fat is a major contributor to the disease, frequently cite epidemiological studies that demonstrate exceptions to the general rule and complain (with some justification) that the adherents to the lipid hypothesis 'conveniently' forget these exceptions. Thus the Israeli diet provides a relatively high fat intake with what might be regarded as a favourably high P/S ratio, but the death rate from heart disease is among the highest in the world. The Scots, with a relatively modest average blood cholesterol value, are now near the top of the world league for heart disease. The Masai of Kenya eat enormous quantities of saturated fats and cholesterol yet have a low incidence of heart disease and incidentally low concentrations of high density lipoproteins (HDL).

The importance of these exceptions falls into perspective if it is borne firmly in mind that blood lipoprotein patterns are *indicators* either of the kind of diet we eat or of deep-seated metabolic changes which may themselves be involved in the cause of the disease or that may have arisen *as a result of* the disease (whose primary cause was entirely different). A blood lipoprotein pattern may be the result of several independent metabolic changes not all of which may be associated with the disease. If epidemiology can only supply clues about diet and disease, is there any direct experimental evidence to support any part of the lipid hypothesis? Feeding high fat diets containing a major proportion of saturated fats, with or without cholesterol, can give rise to arterial lesions that some researchers regard as being sufficiently similar in their pathology to the human lesions to act as a good 'model'. Some animal species develop such lesions when the diet contains only synthetic saturated fats, not cholesterol; others need the presence of cholesterol. Such studies have been useful in understanding the biochemistry of the development of the fatty lesions (i.e. a part of the overall disease process — see Fig. 8.5) but fall short of providing a full model for the human disease because other factors in the human lifestyle, particularly exercise, are often not accounted for in the experiments. Another drawback is that although experimental animals readily develop atherosclerosis, there is no good model for the whole course of the disease, particularly the myocardial infarction. In a sense, one of the best models for studying the relationships between diet, blood lipoproteins and heart disease is the group of people with inherited hyperlipoproteinaemias (See Section 8.3.2). Patients may exhibit a greater than three-fold elevation of serum LDL above 'normal', develop rapidly progressing vascular disease in childhood and show an increased mortality from IHD. Death frequently occurs before the age of 30. That the hyperlipidaemia can be controlled at least in part by diet (drugs are also often needed — see Table 8.6), and thereby life expectancy can be prolonged, is taken by many researchers to indicate that similar measures would be beneficial for those people in the general population whose blood fat concentrations are on the high side of average. The question of dietary strategy will be considered again in the concluding section.

Until fairly recently, attention had been given only to the role of dietary fats in the early stages of cardiovascular disease. In the final stage, involving the formation of a thrombus that may occlude an artery, the adhesiveness of blood platelets is an important factor. During the 1970s it became apparent that the prostaglandins (Section 7.2.3.2) were closely concerned with coagulation in blood vessels and that the balance between the different types of prostaglandins formed may be important in thrombus formation. Prostacyclins (Fig. 7.3) generated in vessel walls inhibit platelet aggregation and this may be the biochemical mechanism whereby the adhesion of platelets to vessel walls is normally prevented. Thromboxane (Fig. 7.3) produced by platelets, has the opposite effect, being a powerful inhibitor of platelet aggregation. The intake of dietary fats, particularly the balance between polyunsaturated and other fatty acids, can affect both the total daily production of prostaglandins and the balance between them. Of particular interest is the observation that a dietary intake of large amounts of docosapentaenoic acid (*n*-3 family, Fig. 2.2(m)), found in some fish oils, can shift the balance towards an anti-aggregating state and could explain the enhanced bleeding tendency in Eskimos, who consume such a diet. Some researchers also believe that it may explain the remarkably low incidence of ischaemic heart disease among Eskimos whose diet, although containing a very high level of fat, is very rich in these highly unsaturated fatty acids from fish tissues. These same scientists have also suggested that the popularization of the role of polyunsaturated fatty acids in the dietary prevention of IHD, with an emphasis on seed oils as a dietary source, has led to a gradual increase in the ratio of *n*-6 to *n*-3 fatty acids in the UK diet. This might adversely affect the coagulating properties of the blood even though it might maintain a useful control of blood lipoprotein concentrations.

This is a relatively new area of research and the relationship between the amount and type of dietary fat, the amounts and types of prostaglandins produced, and the coagulating tendency of the blood is still very obscure. One of the reasons is that there is an enormous range of prostaglandin-like substances that can be produced, even from arachidonic acid alone. These different substances have a multitude of often conflicting physiological effects and much more research is needed before the long term effects of dietary fats on blood coagulation in the living animal are understood and before this knowledge can be translated into practical dietary advice.

In summary, heart disease involves two major phases — atherosclerosis, probably developing over a long time period, and thrombosis, which can occur quite rapidly. It is certain that fats are involved in the biochemistry of both these processes. It is certain that dietary fat can influence the development of atherosclerosis and the tendency for thrombi to form. It is *not* certain whether dietary fat *always* influences either or both of these phases in human beings or the extent to which dietary fat, for an individual, is important in relation to all the other possible risk factors. Blood fat concentrations are indicators that something is happening which may or may not be related to the disease.

Blood fats themselves may or may not be involved as causal agents in the disease.

8.4. NUTRITIONAL MEASURES TAKEN FOR THE PREVENTION OR MANAGEMENT OF DISEASE: DIETARY GUIDELINES: THE ROLE OF FAT

Many people argue that heart disease is such a large public health problem (particularly when it affects a significant number of young people) that any measure which might improve matters should be taken as a matter of urgency even if it cannot be proved that dietary modification can improve the prognosis. Many of these same people also argue that advice about the sort of dietary modification needed to control heart disease should be given to the whole population as part of a national policy. The reasons for this approach are partly that the disease is seen to affect such a large proportion of the population anyway and secondly that the identification of those individuals specifically at risk is impractical as it would require expensive and widespread screening programmes. Others are not inclined to subject whole populations to strictures about what they should or should not eat when it is not known that for any given individual a particular dietary strategy would be effective. It has to be decided whether in any particular case diet is to be seen as a preventative measure, in which case it should be a life-long change, or as a therapeutic measure for those thought to be or known to be at risk. There have been several studies throughout the world designed to discover whether modification of diet can influence the incidence of, or the mortality from heart disease. These so-called 'intervention studies' have been of two kinds. 'Primary' intervention, conducted on apparently healthy subjects over several years during which time the incidence of IHD was observed and 'secondary' intervention, undertaken on patients who have already suffered a 'coronary event', to observe the incidence of re-infarction. They have either concentrated on diet, particularly fat, or like a recent large trial, have examined the effect of modifying 'multiple risk factors'.

Most of the trials involving fat modification have used a relatively low fat diet with a high *P/S* ratio. In general they all achieved some lowering of serum cholesterol and a reduction in the incidence of non-fatal myocardial infarction. Incidence of fatal myocardial infarction was not significantly reduced in any of the trials. Secondary intervention trials

TABLE 8.7
Summary of the Recommendations of the NACNE Discussion Paper

(i) That there should be a standard approach to dietary recommendations for the whole population.

(ii) The choice of average intakes as population goals does not signify that this is the recommendation to which all people should conform. It is easier to make use of epidemiological data to assess the most beneficial average intake of a population than to define the limits of consumption beyond which it is very inadvisable to go.

(iii) Energy intakes should be defined in terms of those appropriate for the maintenance of an optimal bodyweight and adequate exercise. These weights should continue to be defined in terms of height and sex. No increase should be allowed for age, and insurance statistics should continue to be used. With the recent controversy about the need to increase the limits on optimal weights, it needs to be recognised that the *Royal College Report on Obesity* did not advocate changes. The original *Metropolitan Life Insurance Tables* are still appropriate. The public should be encouraged to adjust the types of food eaten, and to increase exercise output so that adult bodyweight is maintained within the optimal limits of weight for height.

(iv) The risk of overweight should not be exaggerated in relation to the risk of continuing to smoke, and this aspect of confusion in the public mind may need to be tackled.

(v) Fat intakes should be on average 30% of total energy intake.

(vi) Saturated fatty acid intake should be on average 10% of total energy intake.

(vii) No specific recommendations should be made on increasing polyunsaturated fatty acids to increase the *P/S* ratio of the diet. The other recommendations will ensure an appreciable increase in this ratio.

(viii) No recommendation is made about lowering cholesterol intake.

(ix) Average sucrose intakes should be reduced to 20 kg per head per year: in calculating the sucrose content of snacks as distinct from total sucrose in the diet, a lower value of 10 kg per head per year should be taken.

(x) Fibre intakes should increase on average to 30 g from 20 g per head per day, the increase to come mainly from the increased consumption of whole grain cereals. An increase in vegetable and fruit consumption should also be advocated.

(xi) It would be desirable if salt intakes on average fell by 3 g per head per day.

(xii) Alcohol intakes should decline to 4% of the total energy intake.

(xiii) Protein intakes should not be altered, but a greater proportion of vegetable protein developing from the other recommendations is appropriate.

TABLE 8.7 (*continued*)

(xiv) Mineral and vitamin intakes which match the recommended allowances listed by the DHSS would be appropriate.

(xv) Special groups need only small additional information, e.g. on the delayed introduction of solids and cereals to infants and the need for dietary vitamin D supplements in some groups of Asian origin. The role of exercise in promoting an increase in total food intake in the elderly is particularly important.

(xvi) Fuller labelling of foods is long overdue, and its health-educational as well as regulatory functions have to be recognised.

A Discussion Paper on Proposals for Nutritional Guidelines for Health Education in Britain, section 11, p. 33, prepared for the National Advisory Committee on Nutrition Education by an *ad hoc* working party under the chairmanship of Professor W. P. T. James. (Health Education Council, London, September 1983.)

have shown little evidence of benefit from a change in diet after a coronary event has been experienced. Such studies are costly, especially if they are to involve significant numbers of subjects to yield statistically valid results and to be carried out over a long enough time period to be effective. However, the longer the period of the study, the more difficult will it be to get subjects to adhere to their diets for the original structure and design of the experiment to be maintained. It is unlikely that there will now be any trial sufficiently well designed to provide proof that any particular dietary regime will be effective in reducing the incidence of heart disease. What measures should be taken is a matter of judgement for the medical profession, for national agencies advising on matters of diet and health and for individuals. If individuals are to choose their diets wisely, they need information and education about foods and their roles in the diet in relation to health.

In the UK, a discussion document has been prepared by a group of nutritionists for the National Advisory Committee on Nutrition Education (NACNE). A summary of its suggestions that relate particularly to dietary fat is presented in Table 8.7. One of the working party's main aims was to express their recommendations in quantitative terms to help health educators in their planning. They do not conceive that it is sensible or possible to achieve these aims in a short period of time but that the recommended changes should be brought about over a 15 year period. Thus, in the 1980s they foresee the average total fat intake being reduced by about 10% from 128 g to about 115 g per head per day and the saturated fat intake by 9 g from 59 g to 50 g per head per day. The ability of the average consumer to comply with these recommendations implies that he or she should be educated about the kinds of foods contributing fat to the diet and that manufactured foods should be labelled in such a way as to allow him or her to make the appropriate choices. It should not be assumed that the apparently simple matter of reducing daily fat intake by 12 g as recommended above is really as straightforward as it seems. We eat food, not fat, carbohydrate, proteins, minerals and vitamins. There may be concomitant nutritional consequences of reducing fat intake depending on which foods are chosen to be reduced.

This point is illustrated below where it is assumed that the 12 g of fat comes entirely from either milk, butter, cheese or beef. Similar points could be made by repeating the exercise with a wide variety of other foods. The milk reduction would involve the removal from the diet of

---

The implications for other nutrients of reducing the intake of energy from fat by 10%.

1. Proportions of energy derived from protein, fat and carbohydrate (figures rounded up or down, from the National Food Survey of 1981 — see MAFF, Bibliography):

Energy source	*National average*		*Dietary goal*	
	% energy	*kJ (kcal) per person per day*	*% energy*	*kJ (kcal) per person per day*
Protein	13	1 400 (325)	13	1 400 (325)
Fat	43	4 500 (1 075)	39	4 050 (970)
Carbohydrate	44	4 600 (1 100)	48	5 050 (1 205)
	100	10 500 (2 500)	100	10 500 (2 500)

Fat goal: reduce by 450 kJ

2. Translating 450 kJ (105 kcal) of fat into food equivalents:
   The above calculation indicates that a 10% reduction of fat is equivalent to 450 kJ (105 kcal). This corresponds to 12 g fat. How might this be translated into foods?

   Table 5.1 indicates the contributions by different foods to the total intake of dietary fat. Major proportions are contributed by meats and by dairy products. We could choose to reduce any one or a combination of foods. Let us suppose we decided to reduce our intake of *either* liquid milk *or* butter or margarine, *or* cheese *or* beef.
   12 g fat are contained in (on average):

315 ml	Milk
14 g	Butter or margarine
42 g	Cheese
150 g	Beef-steak

The removal of these foods (taken individually) from the diet would result in the following losses of nutrients:

Milk	30% loss of recommended daily allowance (RDA) of calcium
	40% loss of RDA for riboflavin
	25% loss of RDA for protein
Butter or margarine	Loss of major contribution of fat-soluble vitamins, especially vitamin A
Cheese	3% loss of RDA for calcium
	20% loss of RDA for protein
Beef-steak	90% loss of RDA for protein
	45% loss of RDA for niacin
	30% loss of RDA for iron
	20% loss of RDA for riboflavin

3. Translating 450 kJ (105 kcal) of carbohydrate into food equivalents:
If energy intake is to stay constant (the NACNE Committee did not think that average energy intakes were too high), then 12 g fat would be exchanged for 28 g carbohydrate. How might this be translated into foods?
28 g carbohydrate would be contained in:

350 g	Porridge
66 g	Wholemeal bread (2 large slices) (includes 5·6 g dietary fibre)
155 g	Potatoes (includes 1·4 g dietary fibre)
150 g	Cooked legumes (lentils, broad beans) (includes 6 g dietary fibre)

These carbohydrate sources would increase dietary fibre consumption as recommended in the NACNE report. However, to increase the palatability of complex carbohydrate it would be usual to add sugar and milk to porridge, butter, jam or cheese to bread, etc. The changes in food patterns would also have important implications for micronutrient intake.

The above example is not given to illustrate the impracticality of dietary recommendations of this sort, but to illustrate that nutritional changes have to be seen in the context of *specific* foods in the context of the *whole* diet. Changes in fat and carbohydrate cannot be discussed in isolation from the foods that contain them and changes in consumption of one food influence the consumption of another. Expert committees issuing guidelines or goals usually overlook this point.

The author is indebted to Dr D. P. Richardson for bringing his attention to the concept on which this example is based, and for many hours stimulating discussion.

40% of the recommended intake of riboflavin. For some people, such reductions could pose a serious problem. Difficulties of this sort can be avoided by choosing foods carefully and eating a wide variety of foods. The main point of this example is to illustrate the importance of considering whole diets, not simply individual foods or worse still, components of foods. In this regard, a great deal of miseducation is achieved by certain types of commercial promotion of margarine or butter in which their respective nutritional (or anti-nutritional) properties are debated. The important factor in good nutrition is the impact of the diet as a whole and each dietary component contributes to good nutrition in so far as it is present in an appropriate amount. It is nonsense, therefore, to speak about the 'good' or 'bad' nutritional qualities of 'butter' or 'margarine' when considered in isolation.

Another conclusion can be drawn from the example given above. It is recommended that the carbohydrate content of the diet is to be increased to balance the energy decrease caused by fat reduction, as the working party suggests that, except for the overweight, average energy intakes need not be reduced. If this were to be achieved by eating more bread, it is unlikely that many people would eat their bread without a fat spread. Therefore, the very act of increasing carbohydrate consumption in certain ways can lead to extra fat being eaten.

The working party believe that there should be a standard approach to dietary recommendations for the whole population; yet in choosing average intakes as population goals, they are careful to point out that this does not signify that these are recommendations to which all people should conform.

In general, people in countries like the UK eat food because they like it, not because they perceive that it will achieve some long term indefinable medical benefit. Those that know that they have a medical problem, or believe that by choosing a diet broadly in line with the recommendations discussed above, may well take a different view. It is the role of nutritionists, dietitians and the health authorities to provide them with information that will guide food choice and it is the role of the food and agricultural industries to make available food products suitably labelled, to satisfy consumer demand. The livestock producer has already responded to a demand for leaner carcass meats. Whether this trend can continue much further is debatable. A consequence of the slower growth that yields a leaner carcass is a relatively more unsaturated and therefore a softer carcass fat which is not desirable to the consumer. The same effect is produced by feeding unsaturated fat supplements (Section 3.3

and Tables 3.1 and 3.5). As well as producing fat of undesirable appearance and texture, an overabundance of unsaturated fat limits the storage life of meat by oxidative breakdown of unsaturated fatty acids leading to peroxide formation and eventually rancidity. The oxidative instability and taste problems associated with products from ruminants fed protected oil seeds was a major factor contributing to the failure of such products becoming established for use in low atherogenic diets. Even when the consumer requires leaner carcasses or modified fat products, the associated changes in colour and softness of the fat may be unacceptable. That is not to say that nutritional needs should not be a stimulus to new product development. In the end, however, there has to be a compromise between nutritional needs, consumer acceptability and the economics of production which should provide a challenge to the agricultural industry.

Given the increasing cost of oils for human foods and animal feeds, and the increasing knowledge of the role of dietary fats, it might be desirable to improve the range and availability of edible oils with desirable characteristics. There are many seed oils with a potentially useful fat composition in species that have hitherto not been used as crops. Interest in the comparative biological activities of different essential fatty acids, for example, has highlighted the possible usefulness of γ-linolenic acid (Section 7.7), leading to the development of evening primrose as a potential agriculture crop (the seed oil contains about 8% of the fatty acids as γ-linolenic acid). Examination of the literature reveals that coconut is not the only seed oil containing medium chain triacylglycerols. *Cuphea* is another provider of these fatty acids that could have agricultural importance. Lupin seeds contain an oil whose major fatty acids are: palmitic, 11%, oleic, 53%; linoleic, 24%; and linolenic, 7%. Either lupins or sunflowers could provide additional oil seed crops in cool climates such as exist in the UK. The latter seed oil is particularly rich in linoleic acid when grown in cool climates.

The plant breeder has developed varieties of rape to eliminate the potentially toxic fatty acid, erucic acid. However, the work of the plant breeder is extremely slow and there is always the possibility that breeding in desired characteristics or breeding out undesired characteristics is accompanied by the elimination of other desirable features such as cold or disease resistance or high yield. Attempts to breed out the high linolenic acid content of soybean oil to improve flavour and oxidative stability have been disappointing. Techniques for growing a variety of plant types in tissue culture have been developed in recent years. Such cultures can be exploited in two main ways: to regenerate clones of plants for use in conventional agriculture or as a source of low-volume, high-cost products such as drugs, food colourings or flavours. During the last decade, the ability to propagate clones of genetically identical plantlets in such commercially important crops as the oil palm has become a reality. Propagation by tissue culture techniques promises to reduce variability in quality and yield and should bring plants into production more quickly, though at a slightly greater cost than conventional methods. Several research groups are now directing their attention to the possibility of inserting into specified plant cells the genes for a desired pathway of fat metabolism. In this way, the fat composition of plant products could be manipulated in any way that was thought to have a nutritional advantage. Such an approach needs not only improved techniques for the genetic manipulation of higher plant cells, which is lagging behind the microbial technology, but also a more detailed knowledge of the genetic control of the pathways of fat metabolism and it will be a long time before it has commercial applications.

In the food manufacturing industry there is scope for the provision of a still wider variety of fat modified products: dairy products with different fat contents, fat spreads, blended or interesterified to achieve a desired fat composition and low energy foods with textural properties indistinguishable from their high fat counterparts. Advances in nutritional knowledge should act as a stimulus for manufacturers to provide an imaginative range of products to suit consumer needs rather than being seen as a 'threat' to parts of the industry as is sometimes now the case.

8.8. SUMMARY

Good nutrition is an important factor in ensuring good health. Much of the world's population suffers from malnutrition which could in part be alleviated by an increase in dietary fats to supply energy, essential fatty acids and vitamin A. Conversely, many persons in affluent countries suffer from obesity and vascular diseases. Reduction in the intake of dietary fats, while ensuring an adequate intake of essential fats, has a part to play in controlling these conditions at least for some individuals. Because of interactions and competition between nutrients, optimum nutrition does not simply depend on the absolute quantities of nutrients supplied, but on the balance between essential and non-essential nutrients in the diet. Thus, some of the signs of essential fatty acid deficiency can occur when an otherwise adequate intake is swamped by an excess of a non-essential fat. Such an imbalance may influence metabolism by changing the properties of biological membranes, thereby affecting cellular function, or by modifying the pattern of production of the prostaglandins. These changes may in turn affect the fine tuning of the immune or the endocrine systems. Subtle modifications of this kind may pave the way, in susceptible individuals, for pathological changes which form the basis for overt disease in later years. Some individuals may have a basic genetic defect resulting in a disorder of fat metabolism that needs to be controlled by appropriate dietary treatment.

Dietary fats, and the balance between them, may thus be involved both in the aetiology of disease or in its nutritional management. So widespread are the cardiovascular diseases, obesity and diabetes that many Western nations are endeavouring to draw up nutritional guidelines for their populations, a common feature of which is a recommendation to reduce the average intake of dietary fats. Now that 70% of all foods eaten in the UK are processed in some way, it is increasingly important for food manufacturers to produce a wide choice of products formulated with both palatability and good nutrition in mind.

BIBLIOGRAPHY

Alexander, J. C., Biological effects due to changes in fats during heating, *J. Am. Oil Chem. Soc.*, 1978, **55**, 711–17.

Artman, N. R., The chemical and biological properties of heated and oxidized fats, *Adv. Lipid Res.*, 1969, **7**, 245.
(Alexander and Artman take different views of the dangers of heated fats).

Beare-Rogers, J. L., *Trans* and positional isomers of common fatty acids, *Advances in Nutritional Research*, 1983, **5**, 171–200. (A detailed up to date account of the structure, occurrence, metabolism and toxicology of the isomeric fatty acids.)

Brisson, G. J., *Lipids in Human Nutrition: An Appraisal of Some Dietary Concepts*, 1981, MTP Press, Lancaster. (Contains a chapter on the nutritional implications of *trans* fatty acids and a chapter warning against developing a 'phobia' about cholesterol.)

Burt, R. and Buss, D. H., Dietary fatty acids in the UK, *J. Clinical Practice*, 1984, in press. (The only detailed up to date analysis of the different kinds of fat eaten in the UK.)

Coates, M. E., Dietary lipids and ischaemic heart disease, *J. Dairy Research*, 1983, **50**, 541–57. (A balanced and readable review of the evidence for the involvement of lipids in arterial disease.)

Dhopeshwarkar, G. A., Naturally occurring food toxicants: toxic lipids, *Prog. Lipid Res.*, 1981, **19**, 107–18. (A detailed review of a variety of lipids found to have toxic properties.)

Kolata, G. B., Atherosclerotic plaques: competing theories guide research, *Science*, 1976, **194**, 592–4.

Kolata, G. B. and Marx, J. L., Epidemiology of heart disease: searches for causes, *Science*, 1976, **194**, 509–12.

MAFF, Household food consumption and expenditure, *Annual Report of the National Food Survey Committee*, 1981, HMSO, London.

Marx, J. L., Atherosclerosis: the cholesterol connection, *Science*, 1976, **194**, 711–14, 755.

(Although now somewhat dated, the papers by Kolata, Kolata & Marx and Marx give an excellent account for the non-specialist, which helps to get some of the controversies into perspective.)

National Advisory Committee on Nutrition Education, *A Discussion Paper on Proposals for Nutritional Guidelines for Health Education in Britain*, 1983, Health Education Council, London. (This is a discussion document, drawn up by an *ad hoc* working party chaired by Professor W. P. T. James and which effectively encapsulates modern nutritional thinking about the 'healthy diet'.)

Oliver, M. F., Diet and coronary heart disease, *Br. Med. Bull.*, 1981, **37**, 49–58. (A short, readable review in which dietary fats are put in perspective with various other possible dietary influences on heart disease.)

Perkins, E. G. and Visek, W. J. (Eds), *Dietary Fats and Health*, 1983, American Oil Chemists' Society, Champaign, Illinois. (A must for the reader who wants to keep up to date with the latest in fats and health in briefly summarized form. Contains chapters on Isomeric Fatty Acids, Heart Disease, Lipoproteinaemias, Cancer. Other subjects dealt with are Essential Fatty Acid Deficiency, Immunity, Aging and Multiple Sclerosis.)

Turner, M. and Gray, J. (Eds), *Implementation of Dietary Guidelines: Obstacles and Opportunities*, 1982, British Nutrition Foundation, London. (Reviews, in a readable way, the whole question of designing guidelines taking account of changing patterns of food consumption and disease incidence in the UK, the individual variations in nutrient requirements and the constraints of food production and supply.)

Vergroesen, A. J. (Ed.), *The Role of Fats in Human Nutrition*, 1975, Academic Press, London. (Contains chapters on Atherosclerosis, Malabsorption, Serum Lipids, Diabetes, Thrombosis and Rapeseed Oil Lipids and Their Biological Effects.)

Index